Advancing Maths for AQA
STATISTICS 1

Roger Williamson, Gill Buqué, Jim Miller and Chris Worth

D.two 8/09 £10.95

Series editors

Roger Williamson   Sam Boardman   Ted Graham
David Pearson

www.heinemann.co.uk
✓ Free online support
✓ Useful weblinks
✓ 24 hour online ordering

01865 888058

Heinemann
Inspiring generations

Heinemann is an imprint of Pearson Education Limited,
a company incorporated in England and Wales, having
its registered office at Edinburgh Gate, Harlow, Essex,
CM20 2JE. Registered company number: 872828

Heinemann is a registered trademark of Pearson Education Limited

© Roger Williamson, Gill Buqué, Jim Miller and Chris Worth 2000, 2004
Complete work © Heinemann Educational Publishers 2004

First published 2004

08
10 9 8

British Library Cataloguing in Publication Data is available from the British
Library on request.

ISBN: 978 0 435513 38 2

Edited by Alex Sharpe, Standard Eight Limited
Typeset and illustrated by Tech-Set Limited, Gateshead, Tyne & Wear
Original illustrations © Pearson Education Limited, 2003
Cover design by Miller, Craig and Cocking Ltd
Printed in China (CTPS/08)

Acknowledgements
The publishers and authors acknowledge the work of the writers, David Cassell,
Ian Hardwick, Mary Rouncefield, David Burghes, Ann Ault and Nigel Price of
the *AEB Mathematics for AQA A-Level Series*, from which some exercises and
examples have been taken.

The publishers' and authors' thanks are due to AQA for permission to
reproduce questions from past examination papers.

The answers have been provided by the authors and are not the responsibility
of the examining board.

Every effort has been made to contact copyright holders of material reproduced
in this book. Any omissions will be rectified in subsequent printings if notice is
given to the publishers.

About this book

This book is one in a series of textbooks designed to provide you with exceptional preparation for AQA's new Advanced GCE Specification B. The series authors are all senior members of the examining team and have prepared the textbooks specifically to support you in studying this course.

Finding your way around

The following are there to help you find your way around when you are studying and revising:

- **edge marks** (shown on the front page) – these help you to get to the right chapter quickly;

- **contents list** – this identifies the individual sections dealing with key syllabus concepts so that you can go straight to the areas that you are looking for;

- **index** – a number in bold type indicates where to find the main entry for that topic.

Key points

Key points are not only summarised at the end of each chapter but are also boxed and highlighted within the text like this:

> A **parameter** is a numerical property of a **population** and a **statistic** is a numerical property of a **sample**.

Exercises and exam questions

Worked examples and carefully graded questions familiarise you with the specification and bring you up to exam standard. Each book contains:

- Worked examples and Worked exam questions to show you how to tackle typical questions; Examiner's tips will also provide guidance;

- Graded exercises, gradually increasing in difficulty up to exam-level questions, which are marked by an [A];

- Test-yourself sections for each chapter so that you can check your understanding of the key aspects of that chapter and identify any sections that you should review;

- Answers to the questions are included at the end of the book.

Contents

CHAPTER 1

Introduction to statistics

Learning objectives

After studying this chapter, you should be able to:
- identify different types of variable and to distinguish between primary and secondary data
- understand the terms population, sample, parameter, statistic
- understand the concept of a simple random sample.

1.1 What is statistics?

> Statistics is about all aspects of dealing with data: how to collect it, how to summarise it numerically, how to present it pictorially and how to draw conclusions from it.

This chapter introduces some statistical terminology you will need to understand.

Statistics deals with events which have more than one possible outcome. If you buy a sandwich, in a canteen, priced at £1.20 and offer the cashier a £5 note you should receive £3.80 in change. This is not statistics as there is (or should be) only one amount of change possible.

If the canteen manager wishes to know how much customers spend when visiting the canteen, this is statistics because different customers spend different amounts.

The quantity which varies – in this case the amount of money – is called a **variable**.

Who uses statistics?
- Car manufacturers: to ensure components meet specification.
- Doctors: to compare the results of different treatments for the same condition.
- Government: to plan provision of schools and health services.
- Scientists: to test their theories.
- Opinion pollsters: to find the public's opinion on local or national issues.

Types of variable

Qualitative variables

There are a number of different types of variable.

> **Qualitative variables** do not have a numerical value. Place of birth, sex of a baby and colour of car are all qualitative variables.

Quantitative variables

> **Quantitative variables** do have a numerical value. They can be **discrete** or **continuous**.

- **Discrete variables** take values which change in steps:

$$\textcircled{0} \quad \textcircled{1} \quad \textcircled{2} \quad \textcircled{3}$$

The number of eggs a hen lays in a week can only take whole number values. **Discrete** means separate – there are no possible values in between.

Variables which are counted, such as the number of cars crossing a bridge in a minute, are discrete, but discrete variables are not limited to whole number values. For example, if a sample of five components is examined, the proportion which fail to meet the specifications is a discrete variable which can take the values 0, 0.2, 0.4, 0.6, 0.8 and 1.

- **Continuous variables** can take any value in an interval. For example, height of a child, length of a component or weight of an apple. Such variables are measured, not counted.

12.423 cm

A lizard could be 12 cm long, or 12.5 cm, or any length in between. In practice length will be measured to a given accuracy, say the nearest millimetre. Only certain values will be possible, but in theory there is no limit to the number of different possible lengths.

Sometimes variables which are strictly discrete may be treated as continuous. Money changes in steps of 1p and so is a discrete variable. However, if you are dealing with hundreds of pounds the steps are so small that it may be treated as a continuous variable.

Primary and secondary data

A vast amount of data on a wide variety of topics is published by the government and other organisations. Much of it appears in publications such as *Monthly Digest of Statistics, Social Trends* and *Annual Abstract of Statistics*. These will provide useful data for many investigations. These data are known as **secondary** data as they were not collected specifically for the investigation. Data which are collected for a specific investigation are known as **primary** data.

EXERCISE IA

1 Classify each of the following variables as either qualitative, discrete quantitative or continuous quantitative:

(a) colours of roses,

(b) numbers of bicycles owned by families in Stockport,

(c) ages of students at a college,

(d) volumes of contents of vinegar bottles,

(e) countries of birth of British citizens,

(f) numbers of strokes to complete a round of golf,

(g) proportions of faulty valves in samples of size ten,

(h) diameters of cricket balls,

(i) prices, in £, of chocolate bars,

(j) makes of car in a car park.

1.2 Populations and samples

What is the average height of women in the UK? To find out, you could, in theory, measure them all. In practice this would be impossible. Fortunately you don't need to. Instead you can measure the heights of a sample. Provided the sample is carefully chosen you can obtain almost as much information from the sample as from measuring the height of every woman in the UK.

In statistics we distinguish between a **population** and a **sample**.

> A **population** is all the possible data and a **sample** is part of the data.

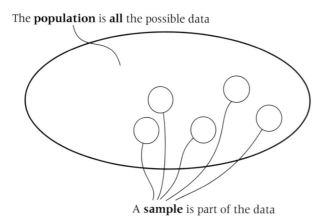

The **population** is **all** the possible data

A **sample** is part of the data

The difference between a population and a sample is of great importance in chapter 6 and in later statistics modules.

Sampling is useful because it reduces the amount of data you need to collect and process. It also allows you to carry out a test without affecting all the population. For example, the contents of a sample of tubs of margarine, from a large batch, might be weighed to ensure that the actual contents matched that claimed on the label. Emptying the tubs to weigh the margarine makes them unsaleable, so it would be ridiculous to weigh the contents of the whole population of tubs.

A numerical property of a population is called a **parameter**. A numerical property of a sample is called a **statistic**. For example, the proportion of tubs in the consignment containing more than 450 g of margarine is a parameter, while the proportion in the sample is a statistic. Similarly, the mean height of all adult females in the UK is a parameter, while the mean height of the adult females in the sample is a statistic.

> A **parameter** is a numerical property of a **population** and a **statistic** is a numerical property of a **sample**.

EXERCISE 1B

1 Read the following passage and identify an example of:

   (a) a population,

   (b) a sample,

   (c) a parameter,

   (d) a statistic,

   (e) a qualitative variable,

   (f) a discrete variable,

   (g) a continuous variable,

   (h) primary data,

   (i) secondary data.

A South American sports journalist intends to write a book about football in his home country. He will analyse all first division matches in the season. He records for each match whether it is a home win, an away win or a draw. He also records for each match the total number of goals scored and the amount of time played before a goal is scored. Reference books showed that in the previous season the mean number of goals per game was 2.317. On the first Saturday of the season he recorded the number of goals scored in each match and calculated the mean number of goals per match as 2.412. For the whole season the mean number of goals per match was 2.219.

EXERCISE 1C

Read the following paragraph:
'All pupils at Gortincham High School undergo a medical examination during their first year at the school. The data recorded for each pupil include place of birth, sex, age (in years and months), height and weight. A summary of the data collected is available on request. A class of statistics students decides to collect data on the weight of second year pupils and compare them with the data on first year pupils. It is agreed that the data will be collected one lunchtime. Each member of the class will be provided with a set of bathroom scales and will weigh as many second year pupils as possible. At the end of the lunchtime they will each report the number of pupils weighed and the mean of the weights recorded.'

1 In the paragraph you have just read identify:
   (a) **two** qualitative variables,
   (b) **two** continuous variables,
   (c) **one** discrete quantitative variable,
   (d) secondary data,
   (e) primary data,
   (f) a population,
   (g) a sample,
   (h) a statistic.

1.3 Sampling without bias

When you are selecting a sample you need to avoid **bias** – anything which makes the sample unrepresentative. For example, if you want to estimate how often residents of Manchester visit the cinema in a year it would be foolish to stand outside a cinema as the audience is coming out and ask people as they pass. This would give a biased sample as all the people you ask would have been to the cinema at least once that year. You can avoid bias by taking a random sample.

Sampling is a major topic in module S2.

Random sampling

For a sample to be random every member of the population must have an equal chance of being selected. However, this alone is not sufficient. If the population consists of 10 000 heights and a random sample of size 20 is required then every possible set of 20 heights must have an equal chance of being chosen.

For example, suppose the population consists of the heights of 100 students in a college and you wish to take a sample of size 5. The students' names are arranged in alphabetical order and numbered 00 to 99. A number between 00 and 19 is selected by lottery methods. For example, place 20 equally sized balls numbered 00 to 19 in a bag and ask a blindfolded assistant to pick one out. This student and every 20th one thereafter are chosen and their heights measured. That is, if the number 13 is selected then the students numbered 13, 33, 53, 73 and 93 are chosen. Every student would have an equal chance of being chosen. However, a sister and brother who were next to each other in the alphabetical list could never both be included in the same sample, so this is **not** a random sample.

Not every set of five students could be chosen.

Usually, if you decide to choose five students at random you intend to choose five different students and would not consider choosing the same student twice. This is known as sampling without replacement.

In this unit you are expected to understand the concept of a simple random sample but it will not be tested in the written examination.

A random sample chosen without replacement is called a **simple random sample**.

Key point summary

1 **Statistics** is about all aspects of dealing with data.   *p1*

2 **Qualitative variables** do not have a numerical value. *p2*

3 **Discrete quantitative variables** take values   *p2*
   which change in steps.

4 **Continuous quantitative variables** can take any   *p2*
   value in an interval.

5 A **population** is all the data.   *p3*

6 A **sample** is part of the data.   *p3*

7 A **parameter** is a numerical property of a   *p4*
   population.

8 A **statistic** is a numerical property of a sample.   *p4*

9 A **random sample** of size *n* is a sample selected
so that all possible samples of size *n* have an
equal chance of being selected.   *p6*

10 **Simple** random samples are selected without
replacement.   *p6*

Test yourself	What to review
1 Explain the difference between a population and a sample.	*Section 1.2*
2 State the type of each of the following variables:	*Section 1.1*
(a) time you have to wait to see a doctor in a casualty department,	
(b) colour of eyes,	
(c) proportion of rainy days in a given week.	
3 Explain the difference between a statistic and a parameter.	*Section 1.2*
4 An inspector tests every 100th assembly coming off a production line. Explain why this is not a random sample of the assemblies.	*Section 1.3*
5 Explain the difference between primary and secondary data.	*Section 1.1*

Test yourself ANSWERS

1 A population is all the data, a sample is part of the data.

2 (a) Continuous quantitative;

    (b) Qualitative;

    (c) Discrete quantitative.

3 A statistic is a numerical property of a sample, a parameter is a
numerical property of a population.

4 Not all combinations of assemblies could be included in the sample. For
example, two adjacent assemblies could not both be sampled.

5 Primary data are data collected specifically for a particular investigation.
Secondary data may be used in the investigation but were not collected
specifically for this purpose.

CHAPTER 2

Numerical measures

Learning objectives

After studying this chapter, you should be able to:
- calculate the mode, median and mean
- calculate the standard deviation, variance, range and interquartile range
- use these statistics to compare sets of data
- select numerical measures which are appropriate in particular circumstances.

You will have met most of the material in this chapter when studying for GCSE. Beware – questions on this topic are often badly answered, probably because students think it is unnecessary to revise these topics.

2.1 Measures of average

There are three main measures of the 'average' of a set of numerical data, which you will have met at GCSE: the mode, the median and the mean.

Low average · · · High average

> These three measures of average tell us something that is typical of a set of data.

Mode

> The mode is the most frequently occurring value (most popular), and is the easiest to obtain – just see which value occurs most often in the data.

> Sometimes the term 'modal value' is used.

Median

> The median is the central value when the data are arranged in order of magnitude.

Mean

> The mean is commonly called the 'average' value, and requires more calculation than either of the other two measures. You add all the values, then divide this total by the number of values.

> Unlike the other 'averages,' all observations contribute to the mean. For most purposes it is the most useful measure of average.

Worked example 2.1

The number of fish caught by each of 20 anglers is given in the table below.

$$4 \quad 7 \quad 12 \quad 13 \quad 0 \quad 5 \quad 21 \quad 13 \quad 10 \quad 6$$
$$6 \quad 8 \quad 15 \quad 9 \quad 6 \quad 0 \quad 14 \quad 12 \quad 6 \quad 8$$

Find:

(a) the modal number of fish per angler,

(b) the median number of fish per angler,

(c) the mean number of fish per angler.

Solution

For both the mode and the median, it is clearer if the numbers are arranged in order:

$$0 \quad 0 \quad 4 \quad 5 \quad 6 \quad 6 \quad 6 \quad 6 \quad 7 \quad 8$$
$$8 \quad 9 \quad 10 \quad 12 \quad 12 \quad 13 \quad 13 \quad 14 \quad 15 \quad 21$$

> With an odd number of values, there is one central value. If there are an even number of values, the median is halfway between the two central values.

(a) The **modal** number of fish is **6**, as 6 occurs more frequently than any other value (it occurs four times).

(b) As there are an even number of values (20) you need to look at the 10th and 11th value in the ordered list. As both values are 8, then the **median** number of fish is **8**.

> If the two central values had been 8 and 9, say, then the median would be the mid-point:
>
>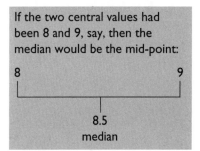

(c) The sum of the 20 values is 175. So the **mean** number of fish is $\dfrac{175}{20} = $ **8.75**.

When the data are given in a frequency table, the calculation of these three averages is made a little easier. The *frequency* is the number of times an observation occurs.

> It is often more convenient to write the frequency table in columns.

Worked example 2.2

A reunion was held 20 years after the members of Form 11 KQ had left the school. The number of children each form member has is given in the table.

Number of children	0	1	2	3	4	5	6
Number of members	9	4	6	5	2	0	1

Calculate **(a)** the mode, **(b)** the median, and **(c)** the mean number of children per member of the form.

No. of children (x)	No. of class members (f)	$(x \times f)$
0	9	0
1	4	4
2	6	12
3	5	15
4	2	8
5	0	0
6	1	6
Total	27	45

Solution

(a) As the most frequent number of children was 0 (nine members had no children), the **mode** is **0**.

(b) There were 27 members at the reunion. The median will be the number of children of the 14th person, when put in order. Nine members had no children, the next four had one child (i.e. these 13 had no children or one child). Therefore, as there are six members in the next group, each having two children, the 14th member must be in this group. The **median** number of children is therefore **2**.

(c) The total number of children of these 27 members is most easily calculated as follows.

$$(0 \times 9) + (1 \times 4) + (2 \times 6) + (3 \times 5) + (4 \times 2) + (5 \times 0) + (6 \times 1) = 0 + 4 + 12 + 15 + 8 + 0 + 6 = 45.$$

The **mean** number of children per member is $\dfrac{45}{27} = 1\frac{2}{3}$ or **1.67** (3 s.f.).

> If the data consists of the whole population the mean is usually denoted by μ. If the data is a sample from the population the mean is denoted $\bar{x}$.

Usually you will be dealing with a sample. In question 2 of exercise 2A the tomato plants are clearly a sample from a population of tomato plants. In this case you can calculate $\bar{x}$, the sample mean. This will almost certainly not be equal to μ, the population mean, but may be used as an estimate of μ. In chapter 6 and in later modules it will be very important for you to be clear about whether you are dealing with a population mean or a sample mean.

> The mean value of a set of data $x_1, x_2, x_3 \dots x_n$ is $\dfrac{\Sigma x_i}{n}$.

The 'Σ' symbol means 'sum of'. If you have a set of values $x_1, x_2, x_3, \dots, x_n$, then

$$\sum_{i=1}^{n} x_i = x_1 + x_2 + x_3 + \dots + x_n.$$

EXERCISE 2A

1 Find the mode, median and mean of these numbers:

$$17, \ 4, \ 9, \ 19, \ 6, \ 11, \ 6, \ 8, \ 9, \ 15, \ 6$$

2 The number of usable tomatoes on 16 tomato plants is given below:

$$8 \quad 5 \quad 20 \quad 17 \quad 10 \quad 9 \quad 12 \quad 9$$
$$15 \quad 11 \quad 9 \quad 10 \quad 19 \quad 10 \quad 9 \quad 14$$

Find the mean, mode and median number of tomatoes per plant.

3 The weights of the eight members of a rowing crew, in kilograms, are 107, 88, 90, 93, 110, 99, 86, 95, to the nearest kilogram.

Find the mean weight of a crew member.

4 The table shows the number of times my team scored 0, 1, 2, ... , goals in their 51 games last season.

Number of goals	0	1	2	3	4	5	6	7
Number of games	7	16	13	7	4	2	0	2

Work out the mode, the median and the mean number of goals per game.

5 A die is thrown 40 times. The scores are shown in the table.

Score	1	2	3	4	5	6
Frequency	4	5	6	10	8	7

Work out **(a)** the mode, **(b)** the median and **(c)** the mean score.

6 In the 24 homework exercises that Henry completed last term, his marks, out of 10 each time, are shown in the table.

Mark (out of 10)	0	1	2	3	4	5	6	7	8	9	10
Frequency	0	0	0	1	3	0	1	1	1	4	13

Work out **(a)** the mode, **(b)** the median, and **(c)** the mean homework mark. Which do you think gives the most realistic mark?

7 A set of five whole numbers has a mode of 3, a median of 4 and a mean of 5. List all possible sets of five numbers having these measures of average.

Discrete and continuous data

The questions in the exercise above all used discrete data – goals in a game, or tomatoes on a plant. You could not have a score of 4.73 on a die, and your team cannot score 2.735 goals in a game. Discrete data can take only certain values (in many cases integers, as in the examples above). You can think of them as **countable** data. If, however, you were to weigh each tomato on a plant, you would not be restricted to whole number answers. Only the accuracy of the weighing scales would limit the number of decimal places you could have. You can think of continuous data as **measurable** – weights, lengths, times, etc. When calculating 'average' values for continuous data, you usually need to put them into suitable groups, or 'class intervals'.

> You have already met discrete and continuous variables in Chapter 1.

Worked example 2.3

The lengths of 50 metal rods are given in metres, correct to the nearest centimetre. Group the lengths into classes, 1.00 m to 1.10 m, 1.10 m to 1.20 m, 1.20 m to 1.30 m, etc, and find the modal length, the median length and the mean length of a rod.

1.34	1.26	1.02	1.53	1.33	1.40	1.19	1.04	1.56	1.44
1.22	1.30	1.13	1.53	1.33	1.40	1.24	1.05	1.24	1.14
1.16	1.32	1.58	1.41	1.25	1.20	1.16	1.13	1.31	1.10
1.44	1.19	1.08	1.47	1.19	1.33	1.13	1.06	1.50	1.32
1.21	1.07	1.22	1.43	1.42	1.03	1.11	1.23	1.33	1.28

> Your calculator is highly likely to have statistical functions where a set of data can be entered, and then the facility of the calculator can give you the mean, variance, standard deviation, and other statistics.
> Some graphical calculators can draw histograms, pie charts, box and whisker plots, scatter diagrams, etc.
> It is important, however, to be aware of the meaning of the various attributes of a set of data and how they are calculated.

Solution

Although it seems straightforward to group the data as requested, we shall need to decide into which group to put a length of, say, 1.20 m. We can decide ourselves, but we need to be consistent. One way (possibly the more conventional way) is to **include** the **lower** boundary and **exclude** the **upper** boundary, in each class interval.

We can write this as '1.00–, 1.10–, 1.20–, …', or '$1.00 \leqslant x < 1.10$, $1.10 \leqslant x < 1.20$, $1.20 \leqslant x < 1.30$', etc.

Putting the lengths into a table will help.

> As the data is given to two decimal places, the grouping asked for in the question cannot be used. This is the nearest you can achieve.

Length of rod	1.00 m –	1.10 m –	1.20 m –	1.30 m –	1.40 m –	1.50 m –
Frequency	7	11	10	9	8	5

The **mode** is **1.33 m**, as it occurs more times (4) than any other length. However, with many different possible lengths, and low frequencies, this is not a useful measure of average. With only small changes you would get a completely different result. For example, if two of the 1.33 m were replaced by 1.32 m and 1.34 m and 1.41 m by 1.44 m, the mode would now be 1.44 m.

The 25th and 26th rod, when put in order, are 1.24 m and 1.25 m. The **median length** is therefore **1.245 m**.

The **mean length** is

$$\frac{(1.34 + 1.26 + 1.02 + \ldots + 1.28)}{50} = \frac{63.35}{50} = \textbf{1.267 m}.$$

Grouped data

If we had not been given individual lengths of rods in example 2.3, but *only* the grouped frequency table, we could not give exact values of the three averages.

Instead of the mode, we could use the **modal class**, which is the '**1.10**–' class, as it has the greatest frequency (11). Compared to the mode in example 2.3, this *is* a useful measure, as it will not be affected by small changes in the data.

See worked example 2.5 if classes are of unequal width.

2

To estimate the mean, we will need to assume that *each* length in a class is equal to the central value of that class – *the class mid-mark* length. As the rods were measured to the nearest centimetre, any rod of length greater than 0.995 m and smaller than 1.095 m would go into the '1.00 m–' class. The mid-mark value of this class is therefore 1.045. Similarly, for the other classes, the mid-mark values are 1.145, 1.245, etc. A table makes it clearer.

Length of rod	1.00 m –	1.10 m –	1.20 m –	1.30 m –	1.40 m –	1.50 m –
Mid value	1.045 m	1.145 m	1.245 m	1.345 m	1.445 m	1.545 m
Frequency	7	11	10	9	8	5

The mean may be obtained directly using your calculator.

The best estimate of the **mean** is:

$$\frac{[(1.045 \times 7) + (1.145 \times 11) + (1.245 \times 10) + (1.345 \times 9) + (1.445 \times 8) + (1.545 \times 5)]}{50} = \textbf{1.275 m}$$

If it had been possible to measure the lengths *exactly*, then the mid-mark values would have been 1.05, 1.15, 1.25, etc. and the mean length would have been 1.28 m.

We could estimate the median like this:

There are 18 rods less than 1.195 m, and there are 10 rods in the next class interval. The 25th and 26th rods will therefore be between the 7th and 8th rods in the group, '1.20 m–'.

Our best **estimate of the median** will be $\dfrac{7\frac{1}{2}}{10}$ into this group,

i.e. $1.195 \text{ m} + \left(\dfrac{7\frac{1}{2}}{10} \times 0.10 \text{ m} \right) = \textbf{1.27 m}$

The median can also be estimated from a cumulative frequency curve.

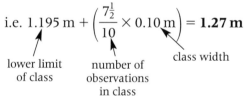

lower limit of class — number of observations in class — class width

Worked example 2.4

Students were asked how long they had spent on solving a homework problem. The results are shown in the table.

The accuracy of the timing is not stated. In this case assume that the times are exact, so the class mid-values are 5, 15, 25, 35, 45 and 55.

Time (*t* min)	$0 \leqslant t < 10$	$10 \leqslant t < 20$	$20 \leqslant t < 30$	$30 \leqslant t < 40$	$40 \leqslant t < 50$	$50 \leqslant t < 60$
No. of students	6	17	20	8	4	0

Estimate the median time and the mean time spent on solving the problem.

Solution

There are 55 students, so the median will be the time spent by the 28th student. This is the 5th student in the $20 \leqslant t < 30$ group. Our estimate of the **median** is $20 + \left(\dfrac{5}{20} \times 10 \right) = \textbf{22.5}$ minutes.

There are $6 + 17 = 23$ times of less than 20 minutes. $28 - 23 = 5$.

Our estimate of the **mean** is:

$$\frac{[(5 \times 6) + (15 \times 17) + (25 \times 20) + (35 \times 8) + (45 \times 4) + (55 \times 0)]}{55}$$

$= 22.636363 \ldots$ or **22.6** minutes.

Worked example 2.5

A company sells clothes by mail order catalogue. The sizes of skirts are defined by the hip measurement; thus customers of the same size will have different heights. The heights, to the nearest centimetre, of a sample of female customers of size 16 were recorded and are summarised in the table below. The sample is thought to be representative of the heights of all female customers of size 16.

Height (cm)	Frequency
131–140	6
141–150	30
151–155	17
156–160	12
161–170	23
171–190	19

(a) For the heights of female customers of size 16;

    (i) identify the modal class,

    (ii) estimate the mean,

    (iii) estimate the median.

(b) The company decides that it is not economic to produce a range of skirts designed for customers of the same size but different heights. For each size, skirts will be designed for customers of a single height. Suggest an appropriate height for customers of size 16. Explain your answer.

Solution

(a) (i) As the classes are of unequal width the modal class is the class with the greatest frequency density. The frequency density is the frequency divided by the class width.

Frequency density is also used when drawing histograms.

Height (cm)	Frequency	Frequency density
131–140	6	0.6
141–150	30	3.0
151–155	17	3.4
156–160	12	2.4
161–170	23	2.3
171–190	19	0.95

As heights have been measured to the nearest centimetre the class 131–140 contains all heights between 130.5 and 140.5 and is therefore of width 10. Similarly the class 151–155 is of width 5, etc.

2

The modal class is 151–155.

(ii)

Height (cm)	Frequency	Class mid-mark
131–140	6	135.5
141–150	30	145.5
151–155	17	153.0
156–160	12	158.0
161–170	23	165.5
171–190	19	180.5

The estimate of the mean may now be found directly using a calculator as 158.0 and this is the recommended method.

If you do not wish to obtain the mean directly using a calculator it is convenient to add a further column (Frequency × Class mid-mark) to the table.

Height (cm)	Frequency, f	Class mid-mark, x	fx
131–140	6	135.5	813
141–150	30	145.5	4365
151–155	17	153.0	2601
156–160	12	158.0	1896
161–170	23	165.5	3806.5
171–190	19	180.5	3429.5
	$\Sigma f = 107$		$\Sigma fx = 16\,911$

$$\text{mean } x = \frac{\Sigma fx}{\Sigma f} = \frac{169\,11}{10} = 158.0$$

(iii) There are 107 observations.

The median will be the $\dfrac{(107+1)}{2} = 54$th in order of magnitude. There are $6 + 30 + 17 = 53$ observations less than 155.5 cm. The median is therefore the $54 - 53 = 1$st observation in the class 156–160. This is estimated by $155.5 + (\frac{1}{12}) \times 5 = 155.9$ cm.

(b) If skirts are to be designed for customers of a single height, the most suitable height to choose is the mode as there are more customers of this than of any other height.

Because height is continuous, a modal class of 151–155, rather than a unique mode has been found. A suitable height to design for would be the mid-point of this class, i.e. 153 cm.

EXERCISE 2B

In the following questions assume all the measurements are exact.

1 The weights of a sample of 40 items taken at random from a production line are given in the table.

Weight (g)	88–	90–	92–	94–	96–	98–100
Frequency	4	6	12	10	6	2

Calculate an estimate of (**a**) the median, and (**b**) the mean weight.

2 An athlete has kept a record of the times taken to win a number of 400 m races, over a period of 18 months.

Time, t (s)	58–	59–	60–	61–	62–	63–	64–	65–66
Frequency	5	8	12	9	14	17	10	5

(**a**) In which period of 1 s does the median 400 m time lie?

(**b**) Estimate the median 400 m time.

(**c**) Calculate an estimate of the mean 400 m time.

3 A market gardener buys some canes. Their lengths are given in the table.

Length (m)	70–	80–	90–	100–	110–	120–	130–140
Frequency	13	27	36	68	51	30	12

(**a**) What is the modal group?

(**b**) Estimate the median length.

(**c**) The mean length is supposed to be within 5 cm of 100 cm. Are these canes within the tolerance allowed?

4 The monthly council tax bills payable by the inhabitants of a village are given in the table.

Amount (£)	140–	160–	180–	200–	220–	240–	260–
Frequency	42	37	24	19	7	12	0

Estimate (**a**) the median, and (**b**) the mean council tax bill.

5 A group of students were asked to work out the answer to a calculation. The number of seconds taken is shown in the table.

Time, t (s)	$20 \leqslant t < 40$	$40 \leqslant t < 60$	$60 \leqslant t < 80$	$80 \leqslant t < 100$
No. of students	4	18	25	13

Estimate the mean time taken to do the calculation.

2.2 Measures of spread

Just as with average values in section 2.1, there are three common measures of spread. The range, the interquartile range, and the standard deviation.

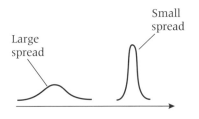

Range

Even though two sets of data have the same mean value, the data can be more 'spread out' in one of these sets, compared to the other. For example, if A = {4, 5, 5, 6, 7, 9} and B = {1, 3, 3, 5, 6, 8, 10, 12}, the mean value of each set is 6. Set A values range from 4 to 9 – a **range** of **5** – but set B has a wider **range** of **11** (from 1 to 12).

> The **range** of a set of data is the **difference** between the **highest** value and the **lowest** value. It is very easy to calculate; all you need are two values – the highest and the lowest.

Interquartile range

When a set of data is written in order of magnitude, the median is the $\frac{(n+1)}{2}$th item of data. The **quartiles** are found in a similar way, the **lower quartile** is the median of the **lower half** of the data, and the **upper quartile** is the median of the **upper half** of the data.

> The range between the quartiles is called the interquartile range ($Q_3 - Q_1$), and covers the middle 50% of the observations.

Worked example 2.6

Find the interquartile range of the following set of data:

    24, 24, 25, 26, 26, 26, 27, 27, 30, 33, 33, 35, 35, 36, 43

Solution

The 15 numbers are already given in numerical order. The median value is the 8th one $\left\{\frac{(15+1)}{2}\text{th}\right\}$, which is 27, the **lower quartile** is the 4th value, which is **26** and the **upper quartile** is the 12th value, which is **35**.

    24, 24, 25, 26, 26, 26, 27, 27, 30, 33, 33, 35, 35, 36, 43
             Q_1 median Q_3

The **interquartile range** is therefore $35 - 26 = $ **9**.

If only a grouped frequency table is given, then you will have to estimate the quartiles, just as for the median above.

The upper quartile of a population is the value which is exceeded by the largest quarter of the observations. Similarly the lower quartile is the value which is exceeded by the largest three-quarters of the observations. This definition can lead to difficulties since if, for example, the population has six members it cannot be split exactly into quarters. Most populations have a large number of members and this problem is of no importance. There are minor disagreements among statisticians about how to find the quartiles of small populations. The method given opposite is one acceptable method.

There are seven observations less than the median and the **lower quartile** is the median of these. Similarly the **upper quartile** is the median of the seven observations which are greater than the median.

The lower quartile is often denoted by Q_1 and the upper quartile by Q_3.

If there had been 16 observations the median, *m*, would have divided the sample into two sets of eight. The **lower quartile** would have been the median of the eight observations less than *m* and the **upper quartile** would have been the median of the eight observations greater than *m*.

Worked example 2.7

When laying pipes, engineers test the soil for resistivity. The table shows the results of 159 tests.

Resistivity (ohm cm^{-1})	Frequency
400–900	5
900–1500	9
1500–3500	40
3500–8000	45
8000–20000	60

Estimate **(a)** the median, and **(b)** the interquartile range of resistivity.

When dealing with grouped data with a total frequency of n the lower quartile is usually taken to be the $\dfrac{n+1}{4}$ th observation and the upper quartile as the $\dfrac{3(n+1)}{4}$ th observation. This is consistent with the median being the $\dfrac{n+1}{2}$ th observation. If n is large it is equally acceptable to use the $\dfrac{n}{4}$ th and $\dfrac{3n}{4}$ th observations for the quartiles.

Solution

The median is the 80th reading, which is the 26th reading in the 3500–8000 group.

$$\textbf{median} = 3500 + \left(\frac{26}{45}\right)(4500) = \textbf{6100} \text{ ohm cm}^{-1}$$

The lower quartile is the 40th reading, which is the 26th reading in the 1500–3000 group.

$$\textbf{lower quartile} = 1500 + \left(\frac{26}{40}\right)(2000) = \textbf{2800}$$

The upper quartile is the 120th reading, which is the 21st reading in the 8000–20 000 group.

$$\textbf{upper quartile} = 8000 + \left(\frac{21}{60}\right)(12\,000) = \textbf{12\,200}$$

Hence the **interquartile range** is 12 200 − 2800 = **9400** ohm cm^{-1}

There are 5 + 9 + 40 = 54 readings up to 3500, and so the 80th reading is the 80 − 54 = 26th reading in the 3500–8000 group.

EXERCISE 2C

1 During the last week, I noted down the number of telephone calls that I made each day. The figures were: 12, 28, 9, 17, 15, 6 and 11. Find the lower and upper quartiles.

2 Find the interquartile range for the data below:

12.5 14.7 15.2 18.7 21.1 25.0 39.7 42.4 48.7 50.5 58.2

3 The number of registered players for 13 of the teams in a cricket league are:

15, 15, 16, 16, 18, 19, 19, 19, 20, 20, 22, 22, 24

Evaluate the median, and the lower and upper quartile number of players.

4 Find the lower and upper quartiles, and the median, of these test scores:

    4, 13, 10, 5, 8, 17, 15, 15, 9, 12, 15, 18, 20, 11, 16

5 Another two students sat the test in question 4. Their scores were 9 and 14. What are the new median, and the new quartile values?

6 The lengths of the 98 telephone calls that I made in question 1 are given in the table.

Call length, t (min)	$0 \leqslant t < 2$	$2 \leqslant t < 4$	$4 \leqslant t < 6$	$6 \leqslant t < 8$	$8 \leqslant t < 10$
Frequency	21	35	22	14	6

Estimate **(a)** the median, **(b)** the lower and upper quartiles, and **(c)** the interquartile range.

7 The table shows the weekly pay of employees in a company.

Pay (£)	0–	100–	200–	300–	400–	500–
Frequency	4	29	21	38	18	7

Estimate **(a)** the median pay, and **(b)** the interquartile range of pay.

8 At a summer fayre, 100 people guessed the amount of money in a large jar. Their guesses are given in the table.

Amount (£x)	$4 \leqslant x < 5$	$5 \leqslant x < 6$	$6 \leqslant x < 7$	$7 \leqslant x < 8$	$8 \leqslant x < 9$	$9 \leqslant x < 10$
Frequency	16	29	35	14	6	0

(a) What is the modal class?

(b) Estimate the median guess.

(c) What is the interquartile range of guesses?

9 What is the mode, median and interquartile range of the number of letters in each day of the week?

10 How many sets of seven non-negative integers can you find that have an interquartile range of 0, a median of 1 and a mean of 2?

Standard deviation

The range and the interquartile range are not entirely satisfactory measures of spread because not all observations contribute to them. For example, the data set $\{1, 1, 1, 1, 1, 6\}$ has the same range as the rather different data set $\{1, 2, 3, 4, 5, 6\}$. One suggestion would be to find the deviation of each observation from the mean and then to find the mean of these deviations. For example, for a population consisting of the numbers 1, 2, 3, 4, 5 the mean, μ, is equal to 3. The deviations from the mean, $x - \mu$, are shown in the table opposite.

x	$x - \mu$
1	-2
2	-1
3	0
4	1
5	2

Unfortunately the mean of the values of $x - \mu$ is zero. This will always be the case and so this will not provide a measure of spread. Instead the deviations are first squared and then the mean is taken.

The mean is $\frac{10}{5} = 2$.

In order to get back to the same units (cm, kg, etc.) as the original population it is now necessary to take the square root. The standard deviation is $\sqrt{2} = 1.41$

x	$x - \mu$	$(x - \mu)^2$
1	-2	4
2	-1	1
3	0	0
4	1	1
5	2	4
	$\Sigma(x - \mu)^2 = 10$	

> The standard deviation of the population $x_1, x_2, x_3, \ldots, x_n$ is usually denoted by σ.
>
> $$\sigma = \sqrt{\Sigma(x_i - \mu)^2/n}$$

Sample standard deviation

The calculation of standard deviation above assumes that the data consists of the whole population.

> Usually your data will be a sample from a population. You will not be able to calculate μ, the population mean. You will only be able to calculate $\bar{x}$, the sample mean. For reasons which are beyond this section the population standard deviation, σ, is 'best' estimated by s, where
>
> $$s^2 = \frac{\Sigma(x - \bar{x})^2}{n - 1},$$
>
> and $\bar{x}$ is the sample mean, $\frac{\Sigma x}{n}$.

Calculators with statistics functions give both s and σ. However, they may be labelled in different ways.
Often σ_{n-1} is used for s and σ_n for σ.
You will need to check your particular calculator, so that you can distinguish which to use for a sample and which to use for a population.

σ = **population** standard deviation.
s is an estimate of σ, calculated from a sample.

Practice obtaining the mean and standard deviation directly using your calculator. This will save you a lot of time.

Calculating the standard deviation

The formula for the standard deviation may be rearranged to simplify the calculation. This was important in days before statistical calculators were widely available. Nowadays, standard deviations may be obtained directly using calculators so this rearrangement is largely obsolete. It is included here for completeness.

$$\Sigma(x - \bar{x})^2 = \Sigma(x^2 - 2x\bar{x} + \bar{x}^2) \quad \text{and} \quad \bar{x} = \frac{\Sigma x}{n}$$

$$= \Sigma x^2 - 2\bar{x}\Sigma x + n\bar{x}^2$$

$$= \Sigma x^2 - 2\frac{(\Sigma x)^2}{n} + \frac{n(\Sigma x)^2}{n^2}$$

$$= \Sigma x^2 - \frac{(\Sigma x)^2}{n}$$

Hence $s^2 = \dfrac{\Sigma(x - \bar{x})^2}{(n - 1)} = \dfrac{\left[\Sigma x^2 - \dfrac{(\Sigma x)^2}{n}\right]}{(n - 1)}$.

This can also be written in the form $\left[\dfrac{\Sigma x^2}{n} - \bar{x}^2\right] \times \dfrac{n}{(n - 1)}$.

Worked example 2.8

The number of travellers queuing to buy tickets at a railway station was observed at random times during the day with the following results:

6 4 1 0 3 8 2 10 0

Calculate the standard deviation:

(a) using the formula $\sqrt{\dfrac{\Sigma(x - \bar{x})^2}{(n-1)}}$,

(b) using the formula $\sqrt{\dfrac{\Sigma x^2 - \dfrac{(\Sigma x)^2}{n}}{(n-1)}}$.

Solution

(a)

x	$x - \bar{x}$	$(x - \bar{x})^2$
6	2.22222	4.93826
4	0.22222	0.04938
1	−2.77778	7.71606
0	−3.77778	14.27162
3	−0.77778	0.60494
8	4.22222	17.82714
2	−1.77778	3.16050
10	6.22222	38.71602
0	−3.77778	14.27162
$\Sigma x = 34$		$\Sigma(x - \bar{x})^2 = 101.55554$

$$\bar{x} = \frac{34}{9} = 3.77778$$

$$s = \sqrt{\frac{101.55554}{(9-1)}} = 3.56$$

(b)

x	x^2
6	36
4	16
1	1
0	0
3	9
8	64
2	4
10	100
0	0
$\Sigma x = 34$	$\Sigma x^2 = 230$

$$s = \frac{\left(230 - \dfrac{34^2}{9}\right)}{(9-1)} = 3.56$$

Now check the answer by finding *s* directly using your calculator.

EXERCISE 2D

In this exercise, assume that the data are *samples*.

1 Calculate the standard deviation of the following data:

6   8   8   9   14   15.

2 Compare the means and standard deviations of the following two sets of data:

A = {3, 4, 5, 6, 7} and B = {1, 3, 5, 7, 9}

3 The number of shots taken by a golfer in the last 12 rounds played is given below.

75,   81,   82,   76,   79,   86,   90,   74,   78,   82,   80,   77.

Calculate the mean score, and the standard deviation.

4 Calculate the standard deviation of the following data:

16 18 18 19 24 25.

Compare your answer with question 1, and explain it.

5 The weight of each member of a rowing crew was measured to the nearest 0.1 kg.

107.3,   87.7,   90.2,   93.0,   109.6,   98.8,   86.4,   95.2.

Calculate the standard deviation of their weights.

6 A class of 15 students scored these marks in a module test:

83, 38, 65, 93, 73, 45, 60, 53, 28, 83, 72, 50, 48, 42, 70

Calculate the mean mark, and the standard deviation.

7 In an experiment, 20 students estimated what they thought was a time interval of 1 minute. Their estimates, in seconds, are shown below.

68, 54, 57, 42, 48, 46, 52, 53, 50, 50,
64, 56, 60, 49, 52, 62, 40, 73, 55, 61.

Calculate the mean estimate, and the standard deviation.

Grouped data

Just as in worked example 2.2, when we calculated the mean value from a frequency table, a convenient way of calculating the standard deviation of grouped data is also to use a table.

> However, it is quicker to obtain this directly from your calculator.

Worked example 2.9

Find the standard deviation of the number of children per class member, from the following data (from worked example 2.2):

Number of children	0	1	2	3	4	5	6
Number of members	9	4	6	5	2	0	1

> This includes all class members so can be regarded as a population.

Solution

Writing the information in columns:

No. of children (x)	No. of class members (f)	x × f	x² × f
0	9	0	0
1	4	4	4
2	6	12	24
3	5	15	45
4	2	8	32
5	0	0	0
6	1	6	36
	27	45	141

The mean value is $\dfrac{45}{27} = 1.66\dot{6}$

$$\text{(standard deviation)}^2 = \left\{\frac{\Sigma(x^2 \times f)}{\Sigma f}\right\} - \mu^2$$

$$= \frac{141}{27} - (1.66\dot{6})^2$$

$$= 2.44\dot{4}$$

Hence the standard deviation $\sigma = \sqrt{2.44\dot{4}} = 1.56$.

If the class members are regarded as a sample from a larger population the appropriate calculation would be

$$s^2 = \left[\left\{\frac{\Sigma(x^2 \times f)}{\Sigma f}\right\} - \bar{x}^2\right] \times \left[\frac{(\Sigma f)}{(\Sigma f - 1)}\right].$$

In this case $\left[\dfrac{141}{27} - (1.66\dot{6})^2\right] \times \dfrac{27}{26} = 2.5385.$

$s = \sqrt{2.5385} = 1.59.$

> There is little difference between this value and the population value of 1.56 calculated above. There is only a large difference if the sample is very small.

Worked example 2.10

A small firm wishes to introduce an aptitude test for applicants for assembly work. The test consists of a mechanical puzzle. The assembly workers, currently employed, were asked to complete the puzzle. They were timed to the nearest second and the times taken by 35 of them are shown below.

Time to complete puzzle (s)	Frequency
20–39	6
40–49	8
50–54	7
55–59	5
60–99	9

(a) Estimate the median and the interquartile range of the data.

(b) Calculate estimates of the mean and the standard deviation of the data.

(c) In addition to the data in the table, five other assembly workers completed the puzzle but took so long that their times were not recorded. These times all exceeded 100 s. Estimate the median time to complete the puzzle for all 40 assembly workers.

(d) The firm decides not to offer employment to any applicant who takes longer to complete the puzzle than the average time taken by the assembly workers who took the test.

    (i) State whether you would recommend the median or the mean to be used as a measure of average in these circumstances. Explain your answer.

    (ii) Write down the value of your recommended measure of average.

Solution

(a) There are 35 observations in all. When they are arranged in order of magnitude the median will be the 18th, the lower quartile will be the 9th and the upper quartile will be the 27th. To estimate these values it is helpful to calculate the cumulative frequency.

Time to complete puzzle (s)	Frequency	Cumulative frequency
20–39	6	6
40–49	8	14
50–54	7	21
55–59	5	26
60–99	9	35

> The cumulative frequency is the total number of observations which do not exceed the upper class bound, e.g. there are $6 + 8 = 14$ observations not greater than 49.5.

The estimate of the median is $49.5 + \left(\frac{4}{7}\right) \times 5 = 52.4$ s.

The estimate of the lower quartile is $39.5 + \left(\frac{3}{8}\right) \times 10 = 43.25$.

The estimate of the upper quartile is $59.5 + \left(\frac{1}{9}\right) \times 40 = 63.94$.

The estimate of the interquartile range is $63.94 - 43.25 = 20.7$ s.

(b)

Time to complete puzzle (s)	Class mid-mark	Frequency
20–39	29.5	6
40–49	44.5	8
50–54	52.0	7
55–59	57.0	5
60–99	79.5	9

> If the data are regarded as the whole population, the standard deviation is 17.1. This would be accepted in an examination. However, it is advisable to always use s unless it is clear from the question that σ is required.

Using a calculator estimate the mean as 54.2 s and the standard deviation as 17.4 s.

(c) The median time for the 40 assembly workers will be between the 20th and 21st when arranged in order of magnitude. Since the additional five workers all took longer than the 35 whose times were recorded there is no need to know the actual times in order to estimate the median. Using the cumulative frequency table in **(a)** the median is estimated by

$$49.5 + \left(\frac{6.5}{7}\right) \times 5 = 54.1 \text{ s.}$$

(d) (i) The median – it is not possible to calculate the mean as the times for the slowest five workers are not known. (Even if the five slowest times were known the median would still be a better measure in these circumstances. A few very long times would greatly increase the mean but have little effect on the median.)

> The median for all 40 assembly workers, not just the fastest 35.

   (ii) 54.1 s.

EXERCISE 2E

In this exercise assume the data are samples.

1 A die was thrown 100 times. The scores are summarised in the table.

Score	1	2	3	4	5	6
Frequency	19	14	13	21	17	16

Calculate the mean and standard deviation of the scores.

2 Find the mean and standard deviation of the number of children per family, for the 23 families shown in the table.

Number of children	0	1	2	3	4	5	6
Number of families	7	3	8	3	0	1	1

3 Applicants for a sales job sit a test consisting of five questions. The number of correct answers, x, scored by 50 candidates is shown in the table.

x	0	1	2	3	4	5
Frequency	30	2	4	5	4	5

Find the mean and the standard deviation of x.

4 The blood pressures, in millimetres of mercury, of a group of 20 athletes are shown in the table.

Blood pressure	65–	70–	75–	80–	85–	90–
Frequency	4	3	5	2	6	0

Calculate the mean and standard deviation of these blood pressures.

5 The breaking strength of 200 cables is given in the table.

Breaking strength (kg × 100)	0–	5–	10–	15–	20–	25–
Number of cables	4	58	66	48	24	0

Estimate the mean breaking strength, and the standard deviation.

Variance

> The variance is the standard deviation squared and for a population is usually denoted σ^2.

Variance plays an important role in mathematical statistics and will appear later in this book and in later statistics units. However it is of little practical use as a measure of spread because it is in inappropriate units.

2.3 Change of scale

The weekly wages paid to the employees of a small engineering firm have a mean of £290 and a standard deviation of £42. The union negotiates a rise of £15 per week for each employee.

Since each wage is increased by £15 the mean wage will be increased by £15 to £305. However, the standard deviation measures variability and this is unchanged if all wages are increased by the same amount. Thus the standard deviation of the new wages remains £42.

> The mode and the median are also each increased by £15. The range and the interquartile range are unchanged.

If instead of a flat-rate rise the union had negotiated an increase of 10% for each employee the variability would increase because the higher-paid employees would receive a larger rise than would the lower-paid employees. In this case both the mean and the standard deviation would increase by 10% – the mean to £319 and the standard deviation to £46.20.

> The mode, median, range and interquartile range all increase by 10%.

Worked example 2.11

A sprinkler, designed to extinguish house fires, is activated at high temperatures. A batch of sprinklers is tested and found to be activated at a mean temperature of 72°C with a standard deviation of 3°C.

Find the mean and standard deviation of the temperatures in degrees Fahrenheit.

Solution

To convert Centigrade to Fahrenheit you must first multiply the temperature by 1.8 and then add 32.

For the mean you should do exactly the same.

Hence the mean in degrees Fahrenheit will be
$72 \times 1.8 + 32 = 161.6$.

For the standard deviation you only carry out the multiplication since the addition of 32 will have no effect on the variability.

Hence the standard deviation in degrees Fahrenheit is
$3 \times 1.8 = 5.4$.

EXERCISE 2F

1 Seedlings in a tray have a mean height of 2.3 cm with a standard deviation of 0.4 cm. Find the mean and standard deviation in millimetres (1 cm = 10 mm).

2 If all the seedlings in question 1 increased in height by 1 cm find the new mean and standard deviation.

3 The mean price of a pair of shoes in a particular shop is £63 with a standard deviation of £18. Find the mean and standard deviation of the prices if, in a sale, all pairs of shoes are reduced in price by:

   (a) £12,

   (b) 50%.

   After two days of the sale the remaining unsold pairs of shoes have a mean price of £70 with a standard deviation of £24. It is decided to reduce the price of each pair by £10 and then label them 'half price'. Thus an unsold pair previously priced at £110 would be sold for £50.

   (c) Find the mean and standard deviation of the new selling prices.

4 The apples in a crate have a median weight of 235 g with an interquartile range of 63 g. Find the median and interquartile range in kilograms (1 kg = 1000 g).

5 It is discovered that the scales used to weigh the apples in question 4 recorded the weights inaccurately. Find the correct median and interquartile range if the weight recorded for each apple was:

   (a) 5 g greater than its actual weight,

   (b) 5% greater than its actual weight.

6 The members of a workers cooperative had mean earnings of £11 500 with a standard deviation of £1000 last year. Find the mean and standard deviation of their total earnings for the year if at the end of the year an additional bonus of:

   (a) £900 was paid to each worker,

   (b) 9% of earnings was paid to each worker.

7 The members of an amateur rugby league team attend training sessions during the week. The number of sessions attended by members of the squad in a particular week has a mode of 3 with a range of 4. Each member of the squad agrees to increase their training sessions by one during the next week. If this is achieved, find the mode and range of the number of training sessions that will be attended during the next week.

2.4 Comparing distributions

The main purpose of calculating numerical measures is to provide a numerical summary of a set of data. This is particularly useful when comparing more than one set of data. For example, the following data represents the value, in £, of the weekly orders taken by Kamran, a sales representative working for a pharmaceutical company.

1260	1190	1480	1790	990	2080	1860	1750
1320	1100	1960	2080	2290	930	1800	1780
1350	1220	2200	1570				

> These observations will be a sample of Kamran's sales figures. Use s, not σ, when calculating the standard deviation.

The weekly orders taken by Debbie, another sales representative working for the same company, over the same period are:

1510	1880	1430	1200	1650	1780	1470	1200
1830	1840	1590	1640	1580	1720	1310	1480
1520	1430	1210	1330				

It is difficult to come to any conclusion by just looking at these blocks of data. You should summarise the data by calculating that Kamran has mean orders of £1600 with a standard deviation of £416 and Debbie has mean orders of £1530 with a standard deviation of £215. It is now easy to see that on average Kamran's orders were slightly higher than Debbie's, but Debbie's orders were much less variable from week to week.

> Kamran's mean is higher than Debbie's.
> Kamran's standard deviation is much higher than Debbie's.

EXERCISE 2G

1 The working lives of torch batteries from manufacturer A have mean 630 hours with a standard deviation of 36 hours. Torch batteries from manufacturer B have mean 412 hours with a standard deviation of 98 hours. Compare the manufacturers.

2 A manufacturing firm has two production lines which have to be stopped for repair and/or adjustment from time to time. The number of minutes between recent stoppages is shown below.

Production line 1	257	354	298	122	98	176	234
	342	401	176	86	138	290	120
Production line 2	198	202	164	257	210	213	189
	172	219	234	214	208	188	216

(a) Calculate the mean and standard deviation of stoppage times for each production line.

(b) Compare the two production lines.

3 Calculate the median and interquartile range of each of the production lines in question 2. Compare the two production lines. Have you reached the same conclusion as in question 2?

4 The breaking strength, in kilograms, of a sample of climbing ropes from each of three suppliers was measured. The following table summarises the results.

Supplier	Mean	Standard deviation
A	856	390
B	820	42
C	496	28

(a) Compare the breaking strength of the ropes from the different suppliers.

(b) Which supplier would you recommend to a friend setting out on a climbing holiday?

5 The following data are the most recent weekly sales, in £, of three representatives working for a confectionery company.

Moira	5120	4970	2230	890	3270	2160	660
	5980	4320	2220				
Everton	4440	3980	4370	3990	3000	3420	2990
	3450	2680	2900				
Syra	2340	2220	2500	2280	3010	2690	2400
	2760	2800	2920				

(a) Calculate the mean and standard deviation of the sales of each representative.

(b) Compare the sales of the three representatives.

6 Calculate the median and the interquartile range of the sales of each of the representatives in question 5. Compare the sales of the three representatives. Have you reached the same conclusions as in question 5?

7 A company sells two makes of washing machines, Ace and Champion, and provides free after-sales service.

(a) For Ace washing machines sold in June 1999, the times, in days, from installing a machine to first being called out to deal with a breakdown are summarised in the following table.

Time to first call-out (days)	Frequency
0–	9
100–	19
200–	28
400–	27
800–1200	17

Estimate the median and the interquartile range.

(b) For Champion washing machines sold in June 1999, the median time from installing a machine to first being called out to deal with a breakdown was 505 days and the interquartile range was 710 days. Compare briefly the reliability of Ace and Champion washing machines.

(c) It was later discovered that 28 Ace washing machines sold in June 1999 had been omitted from the table of data. They had been overlooked because the company had not, after 1200 days, been called out to deal with any breakdowns of these 28 machines. Using this additional information:

    (i) modify the estimates you made in **(a)**,

    (ii) state how, if at all, your answer to **(b)** would be changed.

(d) Give a reason why the median and the interquartile range were used in preference to the mean and standard deviation on times to first call-out.

Key point summary

1 The three most common measures of 'average' are the **mean**, **median** and **mode**.   *p8*

2 The most commonly used is the **mean**. A **sample mean** is denoted $\bar{x}$ and a **population mean** is denoted μ.   *p10*

3 The mean is calculated using the formula $\dfrac{\Sigma x}{n}$.   *p10*

4 The three most common measures of spread are the **range**, **interquartile range** and **standard deviation**.   *p17*

5 The most commonly used measure of spread is the standard deviation. A **population standard deviation** is denoted σ.   *p20*

$$\sigma = \sqrt{\frac{\Sigma(x - \mu)^2}{n}}$$

6 The data you deal with will usually be a sample. If *p20*
this is so it is not possible to calculate σ. You should
estimate σ by s.

$$s = \sqrt{\frac{\Sigma(x - \bar{x})^2}{(n - 1)}}$$

7 You should practise obtaining the mean and *p20*
standard deviation directly using your calculator.
This saves a lot of time and is acceptable in the
examination.

8 The standard deviation squared is called the *p26*
variance.

9 If a variable is increased by a constant amount its *p26*
average will be **increased** by this amount but its
spread will be **unchanged**. This applies whichever
measures of average and spread are used.

10 If a variable is multiplied by a constant amount *p26*
both its average and spread will be multiplied by
this amount. This is true whichever measures of
average and spread are used.

Test yourself	What to review

1 The following data are the girths, in metres, of a sample of trees *Sections 2.1 and 2.2*
in a wood:

    2.1 1.8 3.5 0.8 1.9 0.6 4.6 0.7 1.7

Find the median and the interquartile range.

2 A year later the girth of each tree, in question 1, had increased *Section 2.3*
by 5%. Find the new median and interquartile range.

3 Find the mean and standard deviation of the girth of the trees *Sections 2.1 and 2.2*
in question 1.

4 What symbols would you use to denote the mean and *Sections 2.1 and 2.2*
standard deviation you have calculated in question 3. Explain
your answer.

5 Explain why the mode would not be a suitable measure of *Section 2.1*
average for the data in question 1.

6 The following table summarises the times taken by 70 army *Section 2.1*
recruits to complete an obstacle course.

Time in minutes	10–	12–	14–	16–	18–	20–22
Number of recruits	3	14	30	16	5	2

State the modal class.

Test yourself (*continued*)	What to review
7 Calculate the estimates of the median and the interquartile range for the times in question 6.	*Sections 2.1 and 2.2*
8 Calculate estimates of the mean and standard deviation of the times in question 6.	*Sections 2.1 and 2.2*

Test yourself ANSWERS

8 Mean 15.3, s.d. 2.13.

7 Median 15.2, interquartile range 2.73.

6 Modal class is 14–.

5 The data is too sparse, all observations have a frequency of 1.

4 $\bar{x}$ and s because the girths are a sample.

3 Mean 1.97, s.d. 1.33.

2 Median 1.89, interquartile range 2.15.

1 Median 1.8, interquartile range 2.05.

CHAPTER 3

Probability

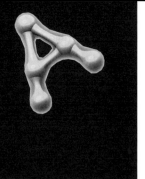

Learning objectives

After studying this chapter, you should be able to:
- understand the concept of probability and be able to allocate probabilities using equally likely outcomes
- identify mutually exclusive events and independent events
- apply the law $P(A \cup B) = P(A) + P(B)$ to mutually exclusive events
- apply the law $P(A \cap B) = P(A)P(B)$ to independent events or the law $P(A \cap B) = P(A)P(B|A)$ to events which are not independent
- solve simple probability problems using tree diagrams or the laws of probability.

3.1 Probability

The concept of probability is widely understood. For example, Courtney might say that the probability of having to wait more than 5 minutes for a bus on a weekday morning is 0.4. He means that if he carries out a large number of **trials** (that is, he waits for a bus on a large number of weekday mornings) he expects to have to wait more than 5 minutes in about 0.4 or 40% of these **trials**.

> Probability is measured on a scale from 0 to 1. Zero represents impossibility and 1 represents certainty.

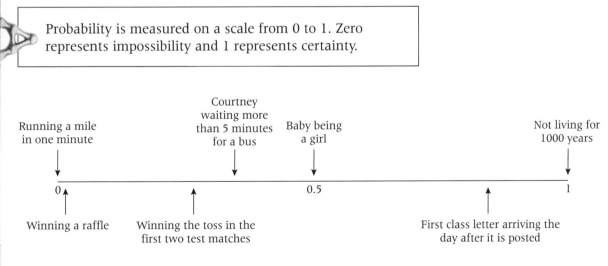

3.2 Equally likely outcomes

Often trials can result in a number of equally likely outcomes. For example, if there are 25 people in a room and a trial consists of choosing one at random there are 25 equally likely outcomes.

> You will understand what is meant by 'equally likely'. Don't try to give it a precise definition – it can't be done.

An **event** consists of one or more of the outcomes. Choosing a particular person, choosing someone wearing glasses or choosing a male would all be examples of events.

If Janice is one of the people in the room then the probability of the event 'choosing Janice' occurring as a result of the trial is $\frac{1}{25}$.

If there are six people wearing glasses in the room the probability of the event 'choosing someone wearing glasses' occurring as a result of the trial is $\frac{6}{25}$.

> If a trial can result in one of n equally likely outcomes and an event consists of r of these outcomes, then the probability of the event happening as a result of the trial is $\frac{r}{n}$.

> It must be impossible for more than one of the outcomes to occur as a result of the same trial.

Worked example 3.1

If a fair die is thrown, what is the probability that it lands showing: **(a)** 2, **(b)** an even number, **(c)** more than 4?

Solution

There are six equally likely outcomes. The probability of any one of them occurring is therefore $\frac{1}{6}$.

(a) $P(2) = \frac{1}{6}$;

(b) $P(\text{even number}) = P(2, 4 \text{ or } 6) = \frac{3}{6} = \frac{1}{2}$;

(c) $P(\text{more than } 4) = P(5 \text{ or } 6) = \frac{1}{3}$.

EXERCISE 3A

1 A box contains 20 counters, numbered 1, 2, 3, ..., 20. A counter is taken out of the box. What is the probability that it: **(a)** is the number 7, **(b)** is a multiple of 4, **(c)** is greater than 14, **(d)** has a 3 on it?

2 The days of the week are written on seven separate cards. One card is chosen. What is the probability that it is: **(a)** Thursday, **(b)** either Monday or Tuesday, **(c)** not Friday, Saturday or Sunday?

3 In a word game, Charlie has the letters B, E, E, H, Q, S and T. One letter accidentally falls on to the floor. What is the probability that it is:

(a) Q, (b) B, E or S, (c) not an E?

4 In a box there are 15 beads. Seven are white, three are yellow, three are blue and two are green. If one bead is selected at random, what is the probability that it is: (a) white, (b) yellow or green, (c) not blue, (d) brown, (e) neither white nor yellow?

5 A cricket team has five batters, a wicketkeeper, three bowlers and two all-rounders. One player is selected at random to pack the cricket bag. What is the probability that the selected player is: (a) the wicketkeeper, (b) a bowler, (c) not an all-rounder, (d) neither a batter nor a bowler?

3.3 Relative frequency

It is not always possible to assign a probability using equally likely outcomes. For example, when Courtney waits for a bus he either has to wait for less than 5 minutes or for 5 or more minutes. However, there is no reason to think that these two outcomes are equally likely. In this case the only way to assign a probability is to carry out the trial a large number of times and to see how often a particular outcome occurs. If Courtney went for a bus on 40 weekday mornings and on 16 of these he had to wait more than 5 minutes he could assign the probability $16/40 = 0.4$ to the event of having to wait more than 5 minutes.

The **relative frequency** of an event is the proportion of times it has been observed to happen.

The **relative frequency** method of assigning probabilities suffers from the problem that if Courtney carried out a large number of further trials it is not possible to prove that the relative frequency would not change completely. However, when this method has been used in practice it has always happened that although the relative frequency may fluctuate over the first few trials, these fluctuations become small after a large number of trials.

In all cases where equally likely outcomes can be used it is also possible to use the relative frequency method. When this has been done it has always been observed that provided a large number of trials are carried out the two methods give very similar although not quite identical results. For example, when a die has been thrown a large number of times the proportion of 1s observed is very close to $\frac{1}{6}$.

> In examination questions you will either be given the probability of an event or be required to find it using equally likely outcomes.

3.4 Mutually exclusive events

When you pick a card from a pack, it must be a club, diamond, heart or a spade. The card cannot be, say, both a club and a heart. The events 'picking a club', 'picking a diamond', 'picking a heart' and 'picking a spade' are said to be **mutually exclusive**. The occurrence of one event **excludes** the possibility that any of the other events could occur.

Worked example 3.2

One card is selected from a pack. Which of these pairs of events are mutually exclusive?

(a) 'The card is a heart' and 'the card is a spade'.

(b) 'The card is a club' and 'the card is a Queen'.

(c) 'The card is black' and 'the card is a diamond'.

(d) 'The card is a King' and 'the card is an Ace'.

(e) 'The card is red' and 'the card is a heart'.

Solution

(a), **(c)** and **(d)**.

In **(b)** it is possible to have a card which is **both** a club **and** a Queen (the Queen of clubs), and in **(e)** a red card could be a heart.

In **(a)**, **(c)** and **(d)** the two events cannot occur simultaneously, hence they are mutually exclusive.

The pack of 52 cards contains 13 clubs, 13 diamonds, 13 hearts and 13 spades. The probability of picking a club is $\frac{13}{52} = 0.25$. The probability of picking a diamond is $\frac{13}{52} = 0.25$. The probability of picking a club or a diamond is $\frac{26}{52} = 0.5$. This is equal to the probability of picking a club plus the probability of picking a diamond. This is an example of the addition law of probability **as it applies to mutually exclusive events**.

> $P(A \cup B)$ denotes the probability of A or B or both occurring. However, here we are dealing with mutually exclusive events so A and B cannot both occur.
>
> If A and B are not mutually exclusive the law is more complicated.

If A and B are mutually exclusive events then the probability of A or B occurring as a result of a trial is the sum of the separate probabilities of A and B occurring as a result of the trial.

$$\mathbf{P}(A \cup B) = \mathbf{P}(A) + \mathbf{P}(B)$$

This law can be extended to more than two mutually exclusive events. For example, the probability of picking a club, a diamond or a heart is $0.25 + 0.25 + 0.25 = 0.75$.

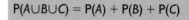

$P(A \cup B \cup C) = P(A) + P(B) + P(C)$

A person, selected at random from 25 people in a room, must either be wearing glasses or not wearing glasses. These two events are mutually exclusive, but one of them must happen. One of these events is called the **complement** of the other. The complement of event A is usually denoted A'.

> Another way of saying that one of the events must happen is to say the events are **exhaustive**.

As one of the events must happen $P(A \cup A') = 1$.

As the events are mutually exclusive
$P(A) + P(A') = P(A \cup A') = 1$

or $P(A') = 1 - P(A)$

> Sometimes it is much easier to work out $P(A)$ than $P(A')$ or vice versa. You can find whichever is easier and, if necessary, use this rule to find the other.

3

EXERCISE 3B

1 The probability of Brian passing a driving test is 0.6. Write down the probability of him not passing the test.

2 The probability of a TV set requiring repair within 1 year is 0.22. Write down the probability of a TV set not requiring repair within 1 year.

3 When Devona rings her mother the probability that the phone is engaged is 0.1, the probability that the phone is not engaged but no one answers is 0.5 and the probability that the phone is answered is 0.4.

Find the probability that:

(a) the phone is engaged or no one answers, **(b)** the phone is engaged or it is answered, **(c)** the phone is not engaged.

4 Kofi shops exactly once a week. In a particular week the probability that he shops on a Monday is 0.3, on a Tuesday is 0.4 and on a Wednesday is 0.1.

Find the probability that he goes shopping on:

(a) Monday or Tuesday, **(b)** Monday or Wednesday,
(c) Monday or Tuesday or Wednesday, **(d)** not on Monday,
(e) Thursday or Friday or Saturday or Sunday.

5 There are 35 customers in a canteen, 12 are aged over 50, 15 are aged between 30 and 50, and five are aged between 25 and 29.

Find the probability that the next customer to be served is aged:

(a) 30 or over, **(b)** 25 or over, **(c)** under 25, **(d)** 50 or under.

6 Charlotte is expecting a baby:

A is the event that the baby will have blue eyes;

B is the event that the baby will have green eyes;

C is the event that the baby will have brown hair.

(a) Write down two of these events which are:

(i) mutually exclusive, **(ii)** not mutually exclusive.

(b) Define the complement of event *A*.

7 A firm employs 20 bricklayers. The Inland Revenue selects one for investigation:

A is the event that the bricklayer selected earned less than £20 000 last year;

B is the event that the bricklayer selected earned more than £20 000 last year;

C is the event that the bricklayer selected earned £20 000 or more last year.

(a) Which event is the complement of *C*?

(b) Are the events *A* and *B* mutually exclusive?

(c) Write down two of the events which are not mutually exclusive.

3.5 Independent events

> When the probability of event *A* occurring is unaffected by whether or not event *B* occurs the two events are said to be **independent**.

For example, if event *A* is throwing an even number with a blue die and event *B* is throwing an odd number with a red die, then the probability of event *A* is $\frac{3}{6} = 0.5$ regardless of whether or not event *B* occurs. Events *A* and *B* are independent.

If two dice are thrown there are 36 equally likely possible outcomes:

1,1	1,2	1,3	1,4	1,5	1,6
2,1	2,2	2,3	2,4	2,5	2,6
3,1	3,2	3,3	3,4	3,5	3,6
4,1	4,2	4,3	4,4	4,5	4,6
5,1	5,2	5,3	5,4	5,5	5,6
6,1	6,2	6,3	6,4	6,5	6,6

If you wish to find the probability of the total score being 12 you can observe that only one of the outcomes (6,6) gives a total score of 12 and so the probability is $\frac{1}{36}$.

If the question concerned more than two dice there would be a very large number of equally likely outcomes and this method would be impractical. An alternative method is to regard throwing a 6 with the first die as event *A* and throwing a 6 with the second die as event *B*. Now use the law that the probability of two independent events both happening is the product of their separate probabilities.

> If *A* and *B* are independent events P($A \cap B$) = P(*A*)P(*B*).

P($A \cap B$) is the probability of events *A* and *B* both happening.

3

The law can be extended to three or more independent events.

The probability of obtaining a total score of 12 (which can only be achieved by throwing a six with both dice) is

$\frac{1}{6} \times \frac{1}{6} = \frac{1}{36}$, as before.

EXERCISE 3C

1 The probability of Brian passing a driving test is 0.6. The probability of Syra passing an advanced motoring test is 0.7. Find the probability of Brian passing a driving test and Syra passing an advanced motoring test.

2 The probability of a TV set requiring repair within 1 year is 0.22. The probability of a washing machine requiring repair within a year is 0.10. Find the probability of a TV set and a washing machine both requiring repair within a year.

3 Two coins are tossed. Find the probability of them both falling heads.

4 The probability that a vinegar bottle filled by a machine contains less than the nominal quantity is 0.1. Find the probability that two bottles, selected at random, both contain:

(a) less than the nominal quantity,

(b) at least the nominal quantity.

5 Three coins are tossed. Find the probability of them all falling tails.

If you wish to find the probability of a total score of 4 when two dice are thrown then you can observe that there are three outcomes which give a total score of 4 and the probability is $\frac{3}{36}$.

1,1	1,2	1,3	1,4	1,5	1,6
2,1	2,2	2,3	2,4	2,5	2,6
3,1	3,2	3,3	3,4	3,5	3,6
4,1	4,2	4,3	4,4	4,5	4,6
5,1	5,2	5,3	5,4	5,5	5,6
6,1	6,2	6,3	6,4	6,5	6,6

Alternatively you can answer the question using the laws of probability.

There are three outcomes which give a total score of 4:

1,3 with probability $\dfrac{1}{6} \times \dfrac{1}{6} = \dfrac{1}{36}$;

2,2 with probability $\dfrac{1}{6} \times \dfrac{1}{6} = \dfrac{1}{36}$;

3,1 with probability $\dfrac{1}{6} \times \dfrac{1}{6} = \dfrac{1}{36}$.

Since these three outcomes are mutually exclusive you can apply the addition law of probability and obtain the probability of obtaining a total score of 4 as $\dfrac{1}{36} + \dfrac{1}{36} + \dfrac{1}{36} = \dfrac{3}{36} = \dfrac{1}{12}$.

Worked example 3.3

The probability that telephone calls to a railway timetable enquiry service are answered is 0.7. If three calls are made find the probability that:

(a) all three are answered,

(b) exactly two are answered.

Solution

If A is the event of a call being answered, $P(A) = 0.7$.

A' is the probability of a call not being answered and $P(A') = 1 - 0.7 = 0.3$.

(a) Using the multiplication law the probability of $AAA = 0.7 \times 0.7 \times 0.7 = 0.343$.

> The law for three independent events is used.

(b) If one call is unanswered it could be the first, second or third call.

$A'AA$ with probability $0.3 \times 0.7 \times 0.7 = 0.147$
$AA'A$ with probability $0.7 \times 0.3 \times 0.7 = 0.147$
AAA' with probability $0.7 \times 0.7 \times 0.3 = 0.147$

> Although A' occurs in different positions the probability of each of the three outcomes is the same.

These three outcomes are mutually exclusive and so you can apply the addition law and find the probability of exactly two calls being answered to be

$0.147 + 0.147 + 0.147 = 0.441$.

3.6 Tree diagrams

An alternative approach to the problem is to illustrate the outcomes with a tree diagram. Each branch shows the possible outcomes of each call and their probabilities. The outcome of the three calls is found by reading along the branches leading to it and the probability of this outcome is found by multiplying the individual probabilities along these branches.

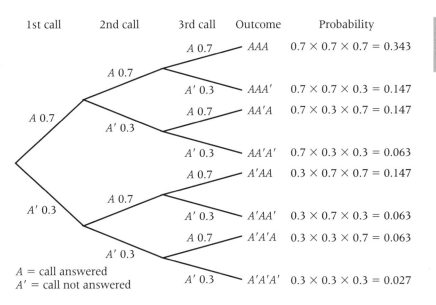

Note the sum of all the probabilities is 1. This is because one of the outcomes must occur.

3

A = call answered
A' = call not answered

The probability of all three calls being answered (*AAA*) can be seen to be 0.343.

The probability of exactly two calls being answered is the sum of the probabilities of the three outcomes *AAA'*, *AA'A* and *A'AA* = 0.147 + 0.147 + 0.147 = 0.441 as before.

Worked example 3.4

A coin is tossed three times. Find the probability that the number of tails is 0, 1, 2 or 3.

Solution

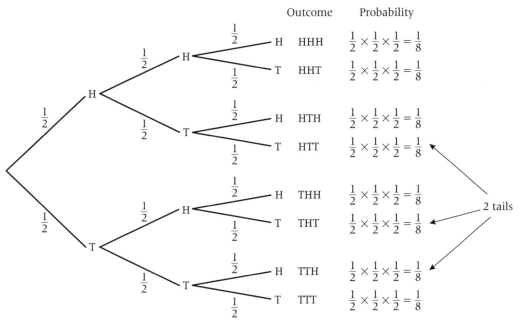

The probability of 0 tails is $\frac{1}{8}$;

The probability of 1 tail is $\frac{1}{8} + \frac{1}{8} + \frac{1}{8} = \frac{3}{8}$;

The probability of 2 tails is $\frac{1}{8} + \frac{1}{8} + \frac{1}{8} = \frac{3}{8}$;

The probability of 3 tails is $\frac{1}{8}$.

EXERCISE 3D

Answer the following questions using tree diagrams or the laws of probability.

1 The probability of Michelle passing a mathematics exam is 0.3 and the probability of her passing a biology exam is independently 0.45.

Find the probability that she:

(a) passes mathematics and fails biology,

(b) passes exactly one of the two examinations,

(c) passes at least one of the two examinations.

2 A civil servant is given the task of calculating pension entitlements. For any given calculation the probability of the result being incorrect is 0.08.

(a) Find the probability that if two pension entitlements are calculated the number incorrect will be:

(i) 0, **(ii)** 1, **(iii)** at least 1.

(b) Find the probability that if three pension entitlements are calculated the number incorrect will be:

(i) 0, **(ii)** 1, **(iii)** 2 or more.

3 The probability of answering a multiple choice question correctly by guessing is 0.25.

(a) A student guesses the answer to two multiple choice questions. Find the probability that:

(i) both answers are correct,

(ii) exactly one answer is correct.

(b) Another student guesses the answer to three multiple choice questions. Find the probability that:

(i) no answers are correct,

(ii) exactly two answers are correct,

(iii) at least two answers are correct,

(iv) less than two answers are correct.

(c) If the second student guesses the answer to four multiple choice questions find the probability that no answers are correct.

4 An opinion poll is to investigate whether estate agents' earnings are thought to be too high, about right or too low. The probabilities of answers from randomly selected adults are as follows:

Too high	0.80
About right	0.15
Too low	0.05.

(a) Find the probability that if two adults are selected at random they will:

(i) both answer 'too high',

(ii) one answer 'too high' and one answer 'about right',

(iii) both give the same answer,

(iv) neither answer 'too low',

(v) both give different answers.

(b) Find the probability that if three adults are selected at random they will:

(i) all answer 'too high',

(ii) two answer 'too high' and one answers 'about right',

(iii) none answer 'too high',

(iv) all give the same answer,

(v) all give different answers.

3.7 Conditional probability

A room contains 25 people. The table shows the numbers of each sex and whether or not they are wearing glasses.

	Male	Female
Glasses	4	5
No glasses	5	11

A person is selected at random.

F is the event that the person selected is female.

G is the event that the person selected is wearing glasses.

There are a total of nine people wearing glasses so $P(G) = \frac{9}{25}$. However, only four of the nine males are wearing glasses and so for the males the probability of wearing glasses is $\frac{4}{9}$ while for the females the probability is $\frac{5}{16}$. That is, the probability of event G occurring is affected by whether or not event F has occurred. The two events are **not independent**.

The **conditional probability** that the person selected is wearing glasses given that they are female is denoted **P(G|F)**.

P(A|B) denotes the probability that event A happens given that event B happens.

Two events A and B are independent if P(A) = P(A|B).

Worked example 3.5

Students on the first year of a science course at a university take an optional language module. The number of students of each sex choosing each available language is shown below.

	French	German	Russian	Total
Male	17	9	14	40
Female	12	11	7	30
Total	29	20	21	70

A student is selected at random.

M is the event that the student selected is male.

R is the event that the student selected is studying Russian.

Write down the value of:

(a) P(M), **(b)** P(R), **(c)** P(M|R), **(d)** P($M \cap R$), **(e)** P($M \cup R$), **(f)** P(R|M), **(g)** P(M'), **(h)** P(R'), **(i)** P(R|M'), **(j)** P(R'|M), **(k)** P($M' \cap R$), **(l)** P($M \cup R'$).

Solution

(a) There are 40 male students out of a total of 70
$$P(M) = \frac{40}{70} = 0.571.$$

(b) 21 students are studying Russian. $P(R) = \frac{21}{70} = 0.3.$

(c) There are 21 students studying Russian of whom 14 are male. $P(M|R) = \frac{14}{21} = 0.667.$

(d) There are 14 students who are both male and studying Russian. $P(M \cap R) = \frac{14}{70} = 0.2.$

(e) There are $17 + 9 + 14 + 7 = 47$ students who are either male or studying Russian (or both). $P(M \cup R) = \frac{47}{70} = 0.671.$

(f) There are 40 male students of whom 14 are studying Russian. $P(R|M) = 14/40 = 0.35$.

(g) There are 30 students who are not male (i.e. are female).

$P(M') = \dfrac{30}{70} = 0.429$.

(h) There are $17 + 9 + 12 + 11 = 49$ students who are not studying Russian. $P(R') = \dfrac{49}{70} = 0.7$.

(i) Of the 30 not male (female) students seven are studying Russian. $P(R|M') = \dfrac{7}{30} = 0.233$.

(j) Of the 40 male students $17 + 9 = 26$ are not studying Russian. $P(R'|M) = \dfrac{26}{40} = 0.65$.

(k) There are seven students who are not male (female) and are studying Russian. $P(M'\cap R) = \dfrac{7}{70} = 0.1$.

(l) There are $17 + 9 + 14 + 12 + 11 = 63$ students who are either male or not studying Russian (or both).

$P(M\cup R') = \dfrac{63}{70} = 0.9$.

3.8 Addition law of probability

In section 3.4, we saw that if A and B are mutually exclusive events then

$$P(A\cup B) = P(A) + P(B)$$

A more general form of this expression which applies whether or not A and B are mutually exclusive events is

$$P(A\cup B) = P(A) + P(B) - P(A\cap B)$$

This is known as the addition law of probability.

The diagram illustrates this law. The dots represent the n equally likely outcomes of a trial.

The event A consists of n_A of these outcomes, the event B consists of n_B of these outcomes. n_{AB} of these outcomes are common to both events A and B.

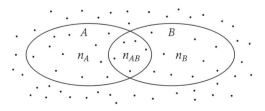

If A and B are mutually exclusive, the event $A \cup B$ would consist of $n_A + n_B$ outcomes. However, if A and B are not mutually exclusive, $n_A + n_B$ would include the n_{AB} outcomes common to both A and B twice.

Hence $A \cup B$ consists of $n_A + n_B - n_{AB}$ outcomes and

$$P(A \cup B) = \frac{(n_A + n_B - n_{AB})}{n} = \frac{n_A}{n} + \frac{n_B}{n} - \frac{n_{AB}}{n} = P(A) + P(B) - P(A \cap B).$$

> You will not be required to prove this result.

Worked example 3.6

One hundred and fifty students at a large catering college have to choose between three tasks as part of their final assessment. The tasks involve cake baking, pastry skills or sweet making. A summary of their choices is given below.

	Male	Female
Cake baking	54	26
Pastry skills	27	18
Sweet making	9	16

A student is selected at random from those at the catering college. C denotes the event that the selected student chooses cake baking. D denotes the event that the selected student is female. Verify that $P(C \cup D) = P(C) + P(D) - P(C \cap D)$.

Solution

There are $54 + 26 = 80$ cake-baking students.

$$P(C) = \frac{80}{150}$$

There are $26 + 18 + 16 = 60$ female students.

$$P(D) = \frac{60}{150}$$

There are 26 female cake-baking students.

$$P(C \cap D) = \frac{26}{150}$$

$$P(C) + P(D) - P(C \cap D) = \frac{80}{150} + \frac{60}{150} - \frac{26}{150}$$
$$= \frac{114}{150}$$

There are $54 + 26 + 18 + 16 = 114$ students who are either cake-baking or female or both.

$$P(C \cup D) = \frac{114}{150} = P(C) + P(D) - P(C \cap D)$$

EXERCISE 3E

1 One hundred and twenty students register for a foundation course. At the end of a year they are recorded as pass or fail. A summary of the results, classified by age, is shown.

	Age (years)	
	Under 20	**20 and over**
Pass	47	33
Fail	28	12

A student is selected at random from the list of those who registered for the course.

Q denotes the event that the selected student is under 20.

R denotes the event that the selected student passed.

(Q' and R' denote the events 'not Q' and 'not R', respectively).

Determine the value of:

(a) $P(Q)$, **(b)** $P(R)$, **(c)** $P(Q')$, **(d)** $P(Q \cap R)$, **(e)** $P(Q \cup R)$,

(f) $P(Q|R)$, **(g)** $P(R'|Q)$, **(h)** $P(Q|R')$, **(i)** $P(Q'|R)$, **(j)** $P(Q \cap R')$.

2 Last year the employees of a firm either received no pay rise, a small pay rise or a large pay rise. The following table shows the number in each category, classified by whether they were weekly paid or monthly paid.

	No pay rise	**Small pay rise**	**Large pay rise**
Weekly paid	25	85	5
Monthly paid	4	8	23

A tax inspector decides to investigate the tax affairs of an employee selected at random.

D is the event that a weekly paid employee is selected.

E is the event that an employee who received no pay rise is selected.

D' and E' are the events 'not D' and 'not E', respectively.

Find:

(a) $P(D)$, **(b)** $P(E')$, **(c)** $P(D|E)$, **(d)** $P(D \cup E)$, **(e)** $P(E \cap D)$,

(f) $P(D \cap E')$, **(g)** $P(E \cup D')$, **(h)** $P(D|E')$, **(i)** $P(E'|D')$.

(j) Verify that $P(E \cup D) = P(E) + P(D) - P(E \cap D)$.

3 A car hire firm has depots in Falmouth and Tiverton. The cars are classified small, medium or large according to their engine size. The number of cars in each class, based at each depot, is shown in the following table.

	Small	Medium	Large
Falmouth	12	15	13
Tiverton	18	22	10

One of the 90 cars is selected at random for inspection.

A is the event that the selected car is based at Falmouth.

B is the event that the selected car is small.

C is the event that the selected car is large.

A′, *B′* and *C′* are the events 'not *A*', 'not *B*' and 'not *C*', respectively.

Evaluate:

(a) P(*A*), **(b)** P(*B′*), **(c)** P(*A*∪*C*), **(d)** P(*A*∩*B*), **(e)** P(*A*∪*B′*), **(f)** P(*A*|*C*), **(g)** P(*C*|*A′*), **(h)** P(*B′*|*A*), **(i)** P(*B*∩*C*).

(j) By comparing your answers to **(a)** and **(f)** state whether or not the events *A* and *C* are independent.

(k) Verify that P(*A*∪*C*) = P(*A*) + P(*C*) − P(*A*∩*C*).

3.9 Multiplication law

If *A* and *B* are independent events P(*A*∩*B*) = P(*A*)P(*B*). This is a special case of the more general law that

$$P(A \cap B) = P(A)P(B|A)$$

A and B may consist of different outcomes of the same trial or of outcomes of different trials.

You can verify this using the earlier example:

	Male	Female
Glasses	4	5
No glasses	5	11

A person is selected at random.

F is the event that the person selected is female.

G is the event that the person selected is wearing glasses.

P(*F*∩*G*), the probability that the person selected is a female wearing glasses, is $\frac{5}{25}$ or 0.2.

Here F and G consist of different outcomes of the same trial.

$$P(F) = \frac{16}{25}$$

P(*G*|*F*), the probability that the person selected is wearing glasses given that they are female, is $\frac{5}{16}$.

$$P(F)P(G|F) = \frac{16}{25} \times \frac{5}{16} = 0.2 = P(F \cap G)$$

Worked example 3.7

Sheena buys ten apparently identical oranges. Unknown to her the flesh of two of these oranges is rotten. She selects two of the ten oranges at random and gives them to her grandson. Find the probability that:

(a) both the oranges are rotten,

(b) exactly one of the oranges is rotten.

3

Solution

(a) P(1st rotten∩2nd rotten) = P(1st rotten) × P(2nd rotten | 1st rotten)

$$P(\text{1st rotten}) = \frac{2}{10}$$

There are now only nine oranges left to choose from, of which one is rotten.

$$P(\text{2nd rotten} \mid \text{1st rotten}) = \frac{1}{9}$$

The probability that both oranges are rotten is

$$\frac{2}{10} \times \frac{1}{9} = \frac{1}{45} = 0.0222.$$

(b) There are two outcomes which result in one rotten orange.

1st rotten 2nd OK with probability $\dfrac{2}{10} \times \dfrac{8}{9} = \dfrac{16}{90}$

or 1st OK 2nd rotten with probability $\dfrac{8}{10} \times \dfrac{2}{9} = \dfrac{16}{90}$.

> Note that although the oranges are selected in a different order the probabilities are the same.

Since these outcomes are **mutually exclusive** we may add their probabilities to obtain the probability of exactly one rotten orange = $\dfrac{16}{90} + \dfrac{16}{90} = \dfrac{32}{90} = 0.356.$

You may prefer to solve examples like this using tree diagrams.

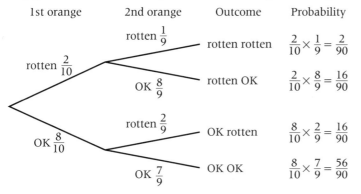

| 1st orange | 2nd orange | Outcome | Probability |

(a) The probability of both oranges being rotten is $\dfrac{2}{90} = 0.0222.$

(b) The probability of exactly one orange being rotten is

$$\frac{16}{90} + \frac{16}{90} = 0.356.$$

Worked example 3.8

When Bali is on holiday she intends to go for a five-mile run before breakfast each day. However, sometimes she stays in bed instead. The probability that she will go for a run on the first morning is 0.7. Thereafter, the probability that she will go for a run is 0.7 if she went for a run on the previous morning and 0.6 if she did not.

Find the probability that on the first three days of the holiday she will go for:

(a) three runs,

(b) exactly two runs. [A]

Solution

(a) Probability Bali goes for three runs (RRR).

P(R 1st morning) $\times$ P(R 2nd morning | R 1st morning) $\times$ P(R 3rd morning | R 2nd morning) $= 0.7 \times 0.7 \times 0.7 = 0.343$.

(b) Probability that Bali runs the first two mornings and stays in bed (B) on the third morning is

P(R 1st morning) $\times$ P(R 2nd morning | R 1st morning) $\times$ P(B 3rd morning | R 2nd morning) $= 0.7 \times 0.7 \times 0.3 = 0.147$.

There are two other possibilities

RBR with probability $0.7 \times 0.3 \times 0.6 = 0.126$

and

BRR with probability $0.3 \times 0.6 \times 0.7 = 0.126$

These outcomes are mutually exclusive and so the probability of Bali going for exactly two runs is

$0.147 + 0.126 + 0.126 = 0.399$.

EXERCISE 3F

1 A car hire firm owns 90 cars, 40 of which are based at Falmouth and the other 50 at Tiverton. If two of the 90 cars are selected at random (without replacement) find the probability that:

(a) both are based at Falmouth,

(b) one is based at Falmouth and the other at Tiverton.

2 Eight students share a house. Five of them own bicycles and three do not. If two of the students are chosen at random to complete a survey on public transport, find the probability that:

(a) both own bicycles,

(b) one owns a bicycle and the other does not.

3 A small firm employs two sales representatives, six administrative staff and four others.

   (a) If two of the 12 staff are selected at random find the probability that they are:
      (i) both sales representatives,
     (ii) one administrator and one sales representative,
    (iii) neither administrative staff.

   (b) If three of the 12 staff are selected at random without replacement find the probability that they:
      (i) are all administrative staff,
     (ii) include two administrative staff and one sales representative,
    (iii) include exactly one sales representative,
    (iv) include one sales representative, one administrator and one other.

4 At a Cornish seaside resort, the local council displays warning flags if conditions are considered to be dangerous for bathing in the sea. The probability that flags are displayed on a particular day is 0.4 if they were displayed on the previous day, and 0.15 if they were not displayed on the previous day.

A family arrives for a three-day holiday. Assume the probability that flags are displayed on their first day is 0.3.

Determine the probability that flags are displayed on:

   (a) all three days,

   (b) none of the three days,

   (c) exactly one of the three days.   [A]

5 The probability of rain interrupting play at a county cricket ground is estimated to be 0.7 if rain has interrupted play on the previous day, and 0.2 if rain has not interrupted play on the previous day.

A three-day match is scheduled to start on Wednesday. The weather forecast suggests that the probability of rain interrupting play on the first day is 0.4.

Find the probability that rain will interrupt play on:

   (a) Wednesday, Thursday and Friday,

   (b) Wednesday and Friday, but not Thursday,

   (c) Thursday and Friday only,

   (d) Friday only,

   (e) exactly one of the first two days (regardless of what happens on Friday),

   (f) exactly one of the three days,

   (g) exactly two of the three days.

6 During an epidemic a doctor is consulted by 40 people who claim to be suffering from flu. Of the 40, 15 are female of whom ten have flu and five do not. Fifteen of the males have flu and the rest do not. If three of the people are selected at random, without replacement, find the probability that:

(a) all three are female,

(b) all three have flu,

(c) all three are females with flu,

(d) all three are of the same sex,

(e) one is a female with flu and the other two are females without flu,

(f) one is male and two are female,

(g) one is a male with flu, one is a male without flu and one is a female with flu.

MIXED EXERCISE

1 A group of three pregnant women attend antenatal classes together. Assuming that each woman is equally likely to give birth on each of the seven days in a week, find the probability that all three give birth:

(a) on a Monday,

(b) on the same day of the week,

(c) on different days of the week,

(d) at a weekend (either a Saturday or Sunday).

(e) How large would the group need to be to make the probability of all the women in the group giving birth on different days of the week less than 0.05? [A]

2 Conveyor-belting for use in a chemical works is tested for strength.

Of the pieces of belting tested at a testing station, 60% comes from supplier A and 40% comes from supplier B. Past experience shows that the probability of passing the strength test is 0.96 for belting from supplier A and 0.89 for belting from supplier B.

(a) Find the probability that a randomly selected piece of belting:

   (i) comes from supplier A and passes the strength test,
   (ii) passes the strength test.

The belting is also tested for safety (this test is based on the amount of heat generated if the belt snaps).

The probability of a piece of belting from supplier A passing the safety test is 0.95 and is independent of the result of the strength test.

(b) Find the probability that a piece of belting from supplier A will pass both the strength and safety tests. [A]

3 Vehicles approaching a crossroad must go in one of three directions – left, right or straight on. Observations by traffic engineers showed that of vehicles approaching from the north, 45% turn left, 20% turn right and 35% go straight on. Assuming that the driver of each vehicle chooses direction independently, what is the probability that of the next three vehicles approaching from the north:

(a) all go straight on,

(b) all go in the same direction,

(c) two turn left and one turns right,

(d) all go in different directions,

(e) exactly two turn left? [A]

4 A bicycle shop stocks racing, touring and mountain bicycles. The following table shows the number of bicycles of each type in stock, together with their price range.

| | Price range | | |
	< £250	£250–£500	> £500
Racing	10	18	22
Touring	36	22	12
Mountain	28	32	20

A bicycle is selected at random for testing.

R is the event that a racing bicycle is selected.

S is the event that a bicycle worth between £250 and £500 is selected.

T is the event that a touring bicycle is selected.

(R', S', T' are the events 'not R', 'not S', 'not T', respectively.)

(a) Write down the value of:
   (i) P(S),   **(ii)** P($R\cap S$),   **(iii)** P($T'\cup S'$)   **(iv)** P($S|R$).

(b) Express in terms of the events that have been defined the event that a mountain bicycle is selected.

(c) Verify that P($R\cup S$) = P(R) + P(S) − P($R\cap S$) [A]

5 The probability that telephone calls to a railway timetable enquiry service are answered is 0.7.

(a) If three calls are made, find the probability that:
   (i) all three are answered,
   (ii) exactly two are answered.

(b) Ahmed requires some timetable information and decides that if his call is not answered he will call repeatedly until he obtains an answer.

Find the probability that to obtain an answer he has to call:
   (i) exactly three times,   **(ii)** at least three times.

(c) If a call is answered, the probability that the information given is correct is 0.8. Thus, there are three possible outcomes for each call:

> call not answered
>
> call answered but incorrect information given
>
> call answered and correct information given.

If three calls are made, find the probability that each outcome occurs once. [A]

6 At the beginning of 1992 a motor insurance company classified its customers as low, medium or high risk. The following table shows the number of customers in each category and whether or not they made a claim during 1992.

	Low	Medium	High
No claim in 1992	4200	5100	3900
Claim in 1992	200	500	1100

(a) A customer is selected at random.

A is the event that the customer made a claim in 1992.

B is the event that the customer was classified low risk.

A' is the event 'not A'.

Write down the value of:
(i) $P(A)$,
(ii) $P(A|B)$,
(iii) $P(B|A')$,
(iv) $P(A \cap B)$,
(v) $P(B \cup A')$.

(b) As a result of the data in **(a)** the company decided not to accept any new high risk customers (but existing customers could continue). In June 1993 its customers were 30% low risk, 65% medium risk and 5% high risk. Use the data in the table above to estimate, for each category of customer, the probability of a claim being made in the next year. Hence estimate the probability that a randomly selected customer will make a claim in the next year. [A]

7 A market researcher wishes to interview residents aged 18 years and over in a small village. The adult population of the village is made up as follows:

Age group	Male	Female
18–29	16	24
30–59	29	21
60 and over	15	25

(a) When one person is selected at random for interview:

A is the event of the person selected being male,

B is the event of the person selected being in the age group 30–59,

C is the event of the person selected being aged 60 or over.

(*A'*, *B'*, *C'* are the events 'not *A*', 'not *B*' and 'not *C*', respectively.)

Write down the value of:

(i) P(*A*), **(ii)** P(*A*∩*B*), **(iii)** P(*A*∪*C'*), **(iv)** P(*B'*|*A*).

(b) When three people are selected for interview, what is the probability that they are all female if:

(i) one is selected at random from each age group,

(ii) they are selected at random, without replacement, from the population of 130 people?

Three people are selected at random, without replacement.

(c) What is the probability that there will be one from each of the three age groups? [A]

Key point summary

1 Probability is measured on a scale from 0 to 1. *p33*

2 If a trial can result in one of *n* equally likely *p34* outcomes and an event consists of *r* of these, then the probability of the event happening as a result of the trial is $\frac{r}{n}$.

3 Two events are **mutually exclusive** if they cannot *p36* both happen.

4 If *A* and *B* are **mutually exclusive** events *p36* P(*A*∪*B*) = P(*A*) + P(*B*).

5 The event of *A* not happening as the result of a trial *p37* is called the **complement** of *A* and is usually denoted *A'*.

6 Two events are **independent** if the probability of *p38* one happening is unaffected by whether or not the other happens.

7 If *A* and *B* are **independent** events *p48* P(*A*∩*B*) = P(*A*)P(*B*).

8 P(*A*|*B*) denotes the probability that event *A* happens *p44* given that event *B* happens.

9 If events *A* and *B* are **independent** P(*A*|*B*) = P(*A*). *p44*

10 P(*A*∪*B*) = P(*A*) + P(*B*) − P(*A*∩*B*). *p45*

11 P(*A*∩*B*) = P(*A*)P(*B*|*A*) = P(*B*)P(*A*|*B*). *p48*

Test yourself	What to review
1 Twelve components include three that are defective. If two components are chosen at random from the 12 find the probability that: **(a)** both are defective, **(b)** exactly one is defective.	*Section 3.9*
2 Under what conditions does $P(R \cap Q) = P(R)P(Q)$?	*Section 3.5*
3 Under what conditions does $P(S \cup T) = P(S) + P(T)$?	*Section 3.4*
4 A student is selected from a class. R is the event that the student is female. Describe the complement of R. How is the complement usually denoted?	*Section 3.4*
5 There are 15 male and 20 female passengers on a tram. Ten of the males and 16 of the females are aged over 25. A ticket inspector selects one of the passengers at random. A is the event that the person selected is female, B is the event that the person selected is over 25. Write down $P(A)$, $P(B)$, $P(A \mid B)$, $P(A \cap B)$ and $P(A \cup B)$. Hence verify that $P(A \cap B) = P(B)P(A \mid B)$. Why is it not possible to apply the law $P(A \cup B) = P(A) + P(B)$ in this case?	*Sections 3.7, 3.8 and 3.9*
6 It is estimated that the probability of a league cricket match ending in a home win is 0.4, an away win is 0.25 and a draw is 0.35. Find the probability that if three games are played, and the results are independent, there will be: **(a)** three home wins, **(b)** exactly one home win, **(c)** one home win, one away win and one draw.	*Sections 3.5 and 3.6*

Test yourself ANSWERS

6 (a) 0.064; **(b)** 0.432; **(c)** 0.21.

A and B are not mutually exclusive.

5 $\dfrac{4}{7}, \dfrac{26}{35}, \dfrac{8}{13}, \dfrac{16}{35}, \dfrac{6}{7}$; $P(B) \times P(A \mid B) = \dfrac{26}{35} \times \dfrac{8}{13} = \dfrac{16}{35} = P(A \cap B)$,

4 Student is male, R'.

3 S and T mutually exclusive events.

2 R and Q independent events.

1 (a) 0.0455; **(b)** 0.409.

CHAPTER 4

Binomial distribution

Learning objectives

After studying this chapter, you should be able to:
- recognise when to use the binomial distribution
- state any assumptions necessary in order to use the binomial distribution
- apply the binomial distribution to a variety of problems.

4.1 Introduction to the binomial distribution

Bicycle, **Bi**ennial, **Bi**nary and **Bi**nomial all start with 'Bi' which generally implies that 'two' of something are involved. In the binomial distribution, the random variable concerned has two possible outcomes.

You have already met the binomial situation in chapter 3, section 3.6, where the results from a fair coin being thrown three times are considered.

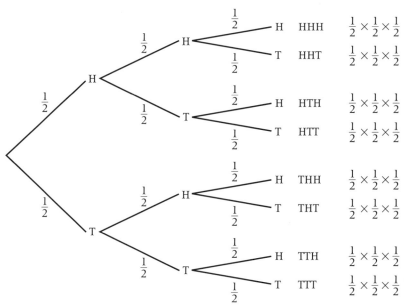

In this situation, the two outcomes are H, a head is showing, or T, a tail is showing.

There are three throws or trials involved and the results which are possible can be written in the following way:

All 3 heads (and 0 tails)	2 heads (and 1 tail)	1 head (and 2 tails)	0 heads (and 3 tails)
HHH	HHT	HTT	TTT
	HTH	THT	
	THH	TTH	
1 way of obtaining all 3 heads	3 ways of obtaining exactly 2 heads	3 ways of obtaining just 1 head	1 way of obtaining no heads

These results are normally written in the following way:

$$P(X = 0) = \left(\frac{1}{2}\right)^3 = \frac{1}{8} \qquad\qquad P(X = 1) = 3 \times \left(\frac{1}{2}\right)^3 = \frac{3}{8}$$

$$P(X = 2) = 3 \times \left(\frac{1}{2}\right)^3 = \frac{3}{8} \qquad\qquad P(X = 3) = \left(\frac{1}{2}\right)^3 = \frac{1}{8}$$

If the coin was not a fair coin but was a coin which had been squashed with a probability of a head showing being $\frac{1}{4}$, rather than $\frac{1}{2}$, then the tree diagram would look like this:

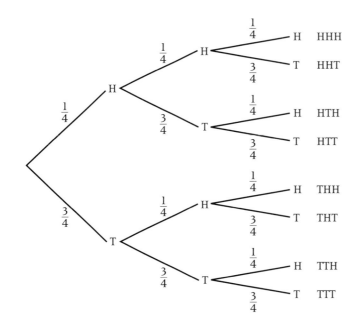

$$\textbf{P(X = 3)} = \left(\frac{1}{4}\right)^3 \qquad\qquad \textbf{P(X = 2)} = 3 \times \left(\frac{1}{4}\right)^2 \times \left(\frac{3}{4}\right)$$

$$\textbf{P(X = 1)} = 3 \times \left(\frac{1}{4}\right) \times \left(\frac{3}{4}\right)^2 \qquad \textbf{P(X = 0)} = \left(\frac{3}{4}\right)^3$$

A tree diagram can be used to find binomial probabilities but only in simple situations.

4.2 The essential elements of the binomial distribution

Certain conditions are necessary for a situation to be modelled by the binomial distribution.

- A fixed number of trials, **n**.

- Just two possible outcomes resulting from each trial.

- The probability of each outcome is the same for each trial.

- The trials are independent of each other.

The letters **n** and **p** are the binomial parameters.

These are often referred to as 'success' and 'failure'.

The probability of a 'success' is called **p**.

Either outcome can be called the 'success' but by convention, you usually refer to the **less** likely outcome as the 'success'.

4

EXERCISE 4A

1 If the probability of a baby being male is 0.5, draw a tree diagram to find the probability that a family of three children has exactly two boys.

2 Four fair coins are thrown. Draw a tree diagram to find the probability that:

   (a) only one head is obtained,

   (b) exactly two heads are obtained.

3 At a birthday party, it is time to light the candles on the cake, but there are only three matches left in the box. If the probability that a match lights is only 0.6, find the probability that none of the matches will light.

4 A regular die has six faces. Two have circles on them and four have squares. This die is rolled three times and the shape on the top of the die is noted.

   Draw a tree diagram to illustrate the possible outcomes and find the probability that:

   (a) none of the rolls resulted in a square shape on the top face,

   (b) at least two throws resulted in a square shape on the top face.

4.3 Investigating further – Pascal's triangle

For larger values of n, a tree diagram is too difficult to construct and a rule is needed to help find the binomial probabilities.

Remember, in the binomial situation:

n represents the number of trials,

p represents the probability of the event concerned.

We can then identify that a random variable X follows a binomial distribution by writing

$$X \sim \mathbf{B}(n, p)$$

In section 4.1

Example 1, $X \sim B\left(3, \dfrac{1}{2}\right)$

Example 2, $X \sim B\left(3, \dfrac{1}{4}\right)$

Worked example 4.1

Ashoke, Theo, Sadie and Paul will each visit the local leisure centre for a swim one afternoon next week. They have not made any particular arrangement between themselves about which afternoon they will go swimming and they are each equally likely to chose any afternoon of the week. The random variable involved, X, is the number of the four friends who go swimming on the Wednesday afternoon of the following week.

Find the probability that exactly two of the friends go swimming on Wednesday afternoon.

Solution

In this situation, $X \sim B\left(4, \dfrac{1}{7}\right)$.

There are four trials as we are considering the number going out of the four friends. This tells us $n = 4$.

They can choose any afternoon therefore $p = \dfrac{1}{7}$.

The tree diagram on the next page shows the probabilities for how many of the friends go on Wednesday. It is very tedious to construct as there are 16 branches.

As the friends are equally likely to choose any afternoon out of the seven available, the probability that one of them chooses Wednesday is $\dfrac{1}{7}$.

$$P(X = 0) = \left(\frac{6}{7}\right)^4 \qquad P(X = 1) = 4 \times \left(\frac{6}{7}\right)^3 \times \left(\frac{1}{7}\right)$$

$$= \frac{1296}{2401} \qquad\qquad = \frac{864}{2401}$$

$$P(X = 2) = 6 \times \left(\frac{6}{7}\right)^2 \times \left(\frac{1}{7}\right)^2 \qquad P(X = 3) = 4 \times \left(\frac{6}{7}\right) \times \left(\frac{1}{7}\right)^3$$

$$= \frac{216}{2401} \qquad\qquad = \frac{24}{2401}$$

$$P(X = 4) = \left(\frac{1}{7}\right)^4 = \frac{1}{2401}$$

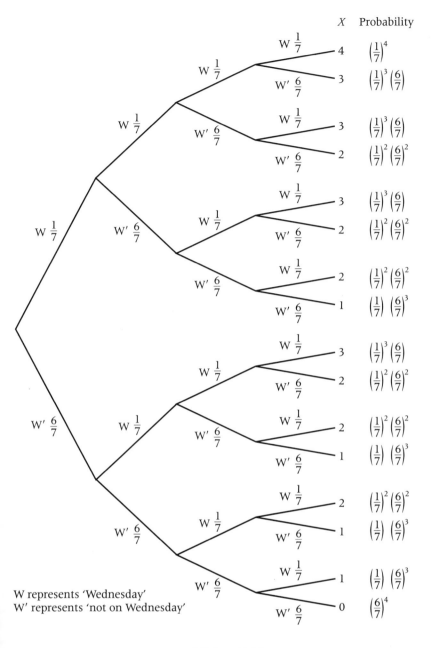

X Probability

W represents 'Wednesday'
W' represents 'not on Wednesday'

Considering **P(X = 2)** $= 6 \times \left(\dfrac{6}{7}\right)^2 \times \left(\dfrac{1}{7}\right)^2$.

The fractions involved are easily explained.

As we are interested in having *two* of the group of four going to the leisure centre on Wednesday and, as the probability for each is $\dfrac{1}{7}$, the $\left(\dfrac{1}{7}\right)^2$ is explained. Also, *two* out of the four will not be going on Wednesday which gives the $\left(\dfrac{6}{7}\right)^2$. The 6 is explained by the fact that there are six branches of the tree diagram which give two $\left(\dfrac{1}{7}\right)$s and two $\left(\dfrac{6}{7}\right)$s.

4

The six ways can be seen by examining the tree diagram but we would not want to draw a tree diagram every time (imagine if there were 10 friends involved!) and so another method needs to be developed to find binomial probabilities.

If we can produce the fractions in the expression for a binomial probability fairly easily, then all that is needed is a way to find the right numbers or coefficients to go with them for all the different ways of combining these probabilities.

In section 4.1, the coefficients for the binomial model where $n = 3$ were found to be 1, 3, 3, 1.

In worked example 4.1, where $n = 4$, the coefficients were found to be 1, 4, 6, 4, 1.

You might recognise these numbers from Pascal's triangle shown below.

Pascal's triangle

					1		1					
				1		2		1				
			1		3		3		1			$\leftarrow n = 3$
		1		4		6		4		1		$\leftarrow n = 4$
	1		5		10		10		5		1	

1   5   10   10   5   1

1   6   15   20   15   6   1

1   7   21   35   35   21   7   1

1   8   28   56   70   56   28   8   1

1   9   36   84   126   126   84   36   9   1

1   10   45   120   210   252   210   120   45   10   1 $\leftarrow n = 10$

> You may well have met Pascal's triangle during GCSE investigations.

Worked example 4.2

Let's imagine now that there are 10 friends, each of whom will visit the leisure centre for a swim one afternoon next week. What is the probability that exactly three of them go for a swim on Wednesday?

> Now, $X \sim B\left(10, \dfrac{1}{7}\right)$.

Solution

It is clearly not possible to use a tree diagram this time but we can start to find the probability by finding the fractions needed.

For three to go on Wednesday, we need $\left(\dfrac{1}{7}\right)^3$. Also, seven must not go on Wednesday which gives $\left(\dfrac{6}{7}\right)^7$.

> There would be 1024 branches for this situation.

We can write down these fractions but how many different ways are there of combining them? According to Pascal's triangle, for the relevant row relating to $n = 10$, there are 120 ways of combining three out of the 10 going on Wednesday and seven not going on Wednesday.

So $P(X = 3) = 120 \times \left(\frac{1}{7}\right)^3 \times \left(\frac{6}{7}\right)^7 = \frac{33\,592\,320}{282\,475\,249}$

$= 0.119$ (three significant figures)

The row is labelled $n = 10$

1	10	45	**120**
$X = 0$	$X = 1$	$X = 2$	**$X = 3$**

Note:
The expressions $\binom{n}{r}$ or $^nC_r = \frac{n!}{r! \times (n-r)!}$.

The expressions $\binom{10}{3}$ or $^{10}C_3 = \frac{10!}{3! \times 7!}$

$= \frac{10 \times 9 \times 8 \times 7 \times 6 \times 5 \times 4 \times 3 \times 2 \times 1}{(3 \times 2 \times 1) \times (7 \times 6 \times 5 \times 4 \times 3 \times 2 \times 1)} = \frac{10 \times 9 \times 8}{(3 \times 2 \times 1)} = 120.$

This is much easier to write down and calculate than drawing 1024 branches!

The number of ways of choosing 3 from 10 is normally written $\binom{10}{3}$ or $^{10}C_3$, so the required probability can be written as

$P(X = 3) = \binom{10}{3} \times \left(\frac{1}{7}\right)^3 \times \left(\frac{6}{7}\right)^7.$

Worked example 4.3

Continuing with the same problem.

Find the probability that exactly half of the friends go to the leisure centre on Wednesday for a swim.

Half means five do go on Wednesday and five do not go.

Solution

The relevant number this time is 252.

The fractions are $\left(\frac{1}{7}\right)^5$ and $\left(\frac{6}{7}\right)^5$

Hence $P(X = 5) = 252 \times \left(\frac{1}{7}\right)^5 \times \left(\frac{6}{7}\right)^5$

$= \frac{1\,959\,552}{282\,475\,249} = 0.00694$ (three significant figures).

Again looking at the row labelled $n = 10$

1	10	45	120	210	**252**
$X = 0$	$X = 1$	$X = 2$	$X = 3$	$X = 4$	**$X = 5$**

$\binom{10}{5}$ or $^{10}C_5 = 252$

Worked example 4.4

Let's extend the problem further to 20 friends, each of whom will visit the leisure centre for a swim one day next week.

What is the probability that exactly five of them will go for a swim on Wednesday?

Solution

This time, both a tree diagram and Pascal's triangle are too difficult to use, but we can easily find the fractions:

Five out of 20 do swim on Wednesday means $\left(\dfrac{1}{7}\right)^5$ and 15 out of 20 do not, which gives $\left(\dfrac{6}{7}\right)^{15}$.

The number of ways of doing this is $\dbinom{20}{5}$ which is 15 504.

> You may find that you need to find $\dbinom{20}{5}$ or $^{20}C_5$ using the ^nC_r button on your calculator.

So $P(X = 5) = 15\,504 \times \left(\dfrac{1}{7}\right)^5 \times \left(\dfrac{6}{7}\right)^{15} = 0.0914$ (three significant figures).

4.4 Finding a formula

When the probability of a 'success' is p and an experiment is repeated n times, then the probability that there are x successes is given by:

> So $X \sim B(n, p)$.

> n is the number of trials.

$$P(X = x) = \binom{n}{x} p^x (1 - p)^{(n - x)}$$

> Use this formula to answer the questions in Exercise 4B.

EXERCISE 4B

1 Ropes produced in a factory are tested to a certain breaking strain. From past experience it is found that one-quarter of ropes break at this strain.

   From a batch of four such ropes, find the probability that exactly two break.

2 A distorted coin, where the probability of a head is $\dfrac{3}{5}$, is thrown five times.

   Find the probability that a head shows on exactly four of these throws.

3 A group of 10 friends plans to each buy a present for their friend who has a birthday. The probability that they will choose to buy chocolates is 0.4 and the friends all choose their present independently.

   Find the probability that only three of the 10 friends decide to buy chocolates.

> In the next section you will see that some binomial probabilities can be obtained from tables. However, the tables do not contain all possible values of a and p. It is essential, for examination purposes, that you are able to calculate binomial probability if necessary.

4 A bank cash dispenser has a probability of 0.2 of being out of order on any one day chosen at random.

Find the probability that, out of the 10 of these machines which this bank owns, exactly three are out of order on any one given day.

Some calculators will give binomial probabilities directly, and this is acceptable in the examination.

5 The probability that Miss Brown will make an error in entering any one set of daily sales data into a database is 0.3.

Find the probability that, during a fortnight (ten working days) she makes an error exactly four times.

6 A restaurant takes bookings for 20 tables on Saturday night. The probability that a party does not turn up for their booking is 0.15.

Find the probability that only two of the parties who have made bookings do not turn up.

4

7 A school pupil attempts a multiple-choice exam paper but has not made any effort to learn any of the information necessary. Therefore the pupil guesses the answers to all the questions.

There are five possible answers to each question and there are 30 questions on the paper. Find the probability that the pupil gets eight questions correct out of the 30 on the paper.

8 A batch of 25 lightbulbs is sent to a small retailer. The probability that a bulb is faulty is 0.1.

Find the probability that only two of the bulbs are faulty.

4.5 Using cumulative binomial tables

We now have a formula to use for evaluating binomial probabilities for precise numbers of successes. We may, however, need to evaluate more than just one individual probability, often a whole series of such probabilities, and this can become very tedious. Fortunately, cumulative binomial probability tables have been constructed to help out in such cases, for certain values of p and n.

Refer to section 4.4.

Tables can be found in the *AQA Formulae Book*, Table I, and at the back of this book.

Worked example 4.5

If an unbiased coin is thrown eight times, what is the probability of obtaining fewer than four heads.

In this case $X \sim B(8, 0.5)$.

Solution

$\overline{0, 1, 2, 3, }4, 5, \ldots$

X **less** than 4

For $P(X < 4)$, we will need

$$P(X = 0) + P(X = 1) + P(X = 2) + P(X = 3)$$
$$= P(X \leqslant 3)$$

This means evaluating and summing four binomial probabilities

This may be found directly on some calculators and is allowed in exams.

$$\left(\frac{1}{2}\right)^8 + \binom{8}{1}\left(\frac{1}{2}\right)^7\left(\frac{1}{2}\right) + \binom{8}{2}\left(\frac{1}{2}\right)^6\left(\frac{1}{2}\right)^2 + \binom{8}{3}\left(\frac{1}{2}\right)^5\left(\frac{1}{2}\right)^3$$

$= 0.3633$ (four decimal places)

but, using the tables, where $n = 8$ and $p = 0.5$, we require $P(X \leqslant 3)$ which is given as a cumulative probability in the tables along the row labelled $x = 3$.

The required probability is found directly as 0.3633.

Tables should always be the first choice for finding any binomial probabilities – they are so much easier. However, not all values of n and p are included so you may need to calculate probabilities.

r	p	0.01	0.02	0.03	0.04	0.05	0.06	0.07	0.08	0.09	0.10	0.15	0.20	0.25	0.30	0.35	0.40	0.45	0.50
$n = 8$	0	0.9227	0.8508	0.7837	0.7214	0.6634	0.6096	0.5596	0.5132	0.4703	0.4305	0.2725	0.1678	0.1001	0.0576	0.0319	0.0168	0.0084	0.0039
	1	0.9973	0.9897	0.9777	0.9619	0.9428	0.9208	0.8965	0.8702	0.8423	0.8131	0.6572	0.5033	0.3671	0.2553	0.1691	0.1064	0.0632	0.0352
	2	0.9999	0.9996	0.9987	0.9969	0.9942	0.9904	0.9853	0.9789	0.9711	0.9619	0.8948	0.7969	0.6785	0.5518	0.4278	0.3154	0.2201	0.1445
	3	1.000	1.000	0.9999	0.9998	0.9996	0.9993	0.9987	0.9978	0.9966	0.9950	0.9786	0.9437	0.8862	0.8059	0.7064	0.5941	0.4770	0.3633
	4			1.000	1.000	1.000	1.000	0.9999	0.9999	0.9997	0.9996	0.9971	0.9896	0.9727	0.9420	0.8939	0.8263	0.7396	0.6367
	5							1.000	1.000	1.000	1.000	0.9998	0.9988	0.9958	0.9887	0.9747	0.9502	0.9115	0.8555
	6											1.000	0.9999	0.9996	0.9987	0.9964	0.9915	0.9819	0.9648
	7												1.000	1.000	0.9999	0.9998	0.9993	0.9983	0.9961
	8														1.000	1.000	1.000	1.000	1.000

The probability of a 'success' should be chosen so that $p \leqslant 0.5$ as found in these tables.

Worked example 4.6

The probability that a candidate will guess the correct answer to a multiple-choice question is 0.2. In a multiple-choice test there are 50 questions. A candidate decides to guess the answers to all the questions and chooses the answers at random, each answer being independent of any other answer.

Find the probability that the candidate:

(a) gets five or fewer answers correct,

(b) gets more than 14 answers correct,

(c) gets exactly nine answers correct,

(d) gets between seven and 12 (inclusive) answers correct.

Solution

In this case, the number of trials, *n* = **50**.

The probability of 'success' (obtaining a correct answer) *p* = **0.2**.

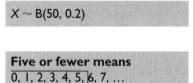

(a) P($X \leq 5$) can be found directly from the tables. As the tables are cumulative, all six probabilities are summed for you

Five or fewer means
0, 1, 2, 3, 4, 5, | 6, 7, ...

$$P(X = 0) + P(X = 1) + P(X = 2) + P(X = 3) + P(X = 4)$$
$$+ P(X = 5) = 0.0480.$$

(b) P($X > 14$) cannot be found directly from the tables but can be easily worked out from a probability that is given there.

more than 14 means
0, 1, 2, ..., 13, 14, | 15, 16, ..., 50
$X \leq 14$ $X > 14$

The opposite event to getting more than 14 correct is getting 14 or fewer correct so we need

$$1 - P(X \leq 14) = 1 - 0.9393$$
$$= 0.0607$$

(c) P($X = 9$) is not given in the tables but, again, it can be found from probabilities that are there by finding

Clearly
0, 1, ..., 7, 8, | 9, | 10, ...
$X \leq 8$
         $X \leq 9$

$$P(X \leq 9) - P(X \leq 8)$$
$$= 0.4437 - 0.3073$$
$$= 0.1364$$

Alternatively, the formula can be used

Look back at section 4.4.

$$P(X = 9) = \binom{50}{9} \times (0.2)^9 \times (0.8)^{41}$$

It is up to you to choose which method you find easier.

$$= 0.1364 \text{ (four decimal places)}.$$

(d) P($7 \leq X \leq 12$) involves summing six individual probabilities which is possible but tedious.

The tables can be used to obtain the answer much more easily by finding

Consider this:
0, ..., 5, 6, | 7, 8, 9, 10, 11, 12, | 13, ..., 50
$X \leq 6$
                 $X \leq 12$

$$P(X \leq 12) - P(X \leq 6)$$
$$= 0.8139 - 0.1034$$
$$= 0.7105$$

Worked example 4.7 ───────────

A golfer practises on a driving range. His aim is to drive a ball to within 20 m of a flag.

The probability that he will achieve this with each particular drive is 0.3.

If he drives 20 balls, what is the probability that he achieves:

(a) five or fewer successes,

(b) seven or more successes,

(c) exactly six successes,

(d) between four and eight (inclusive) successes.

Solution

Clearly, in this situation $n = 20$ and $p = 0.3$.

> $X \sim B(20, 0.3)$

(a) We require $P(X \leq 5) = 0.4164$

> $0, 1, 2, 3, 4, 5, | 6, \ldots, 20$

(b) This time the tables do not supply the answer directly, but it can be found by evaluating

$$1 - P(X \leq 6) = 1 - 0.6080$$
$$= 0.3920$$

> For seven or more successes
> $0, 1, \ldots, 5, 6, | 7, 8, \ldots, 20$
> $\qquad X \leq 6 \qquad$ 7 or more

(c) $P(X = 6)$ can either be found from the tables by evaluating

$$P(X \leq 6) - P(X \leq 5)$$
$$= 0.6080 - 0.4164$$
$$= 0.1916$$

> Clearly
> $0, 1, 2, 3, 4, 5, | 6, | 7, \ldots$
> $\quad X \leq 5$
> $\qquad\qquad X \leq 6$

or from the formula where

$$P(X = 6) = \binom{20}{6} \times (0.3)^6 \times (0.7)^{14}$$
$$= 0.1916$$

(d) $P(4 \leq X \leq 8)$ again can be found from the tables by evaluating

$$P(X \leq 8) - P(X \leq 3)$$
$$= 0.8867 - 0.1071$$
$$= 0.7796$$

> Consider this
> $0, 1, 2, 3, | 4, 5, 6, 7, 8, | 9, 10, \ldots, 20$
> $\quad X \leq 3$
> $\qquad\qquad\qquad X \leq 8$

4.6 The mean and variance of the binomial distribution

If you played ten games of table tennis against an opponent who, from past experience, you know has a chance of only $\frac{1}{5}$ of winning a game against you, how many games do you expect your opponent to win?

> In this example, $X \sim B(10, \frac{1}{5})$.

Most people would instinctively reply 'two games' and would argue that an opponent who wins, on average, $\frac{1}{5}$ of the games, can expect to be successful in $\frac{1}{5}$ of the ten games played. Hence $\frac{1}{5} \times 10 = 2$.

> Mean $= \mu = \frac{1}{5} \times 10 = 2$

In general, if $X \sim B(n, p)$, then the mean of X is given by

> The proof for the results for the mean and the variance are not required but you should be able to quote the results which are given in the AQA formulae book.

> Mean $= \mu = np$

The variance of X is given by

> Variance $= \sigma^2 = np(1 - p)$

In this example, the variance is $10 \times \dfrac{1}{5} \times \dfrac{4}{5} = 1.6$.

Worked example 4.8

A biased die is thrown 300 times and the number of sixes obtained is 80. If this die is then thrown a further 12 times, find:

(a) the probability that a six will occur exactly twice,

(b) the mean number of sixes,

(c) the variance of the number of sixes.

Solution

(a) The probability that a six is obtained is $\dfrac{80}{300} = \dfrac{4}{15}$.

The binomial model in this case is $X \sim \mathrm{B}\left(12, \dfrac{4}{15}\right)$,

so $\mathrm{P}(X = 2) = \dbinom{12}{2}\left(\dfrac{4}{15}\right)^2\left(\dfrac{11}{15}\right)^{10} = 0.211$.

(b) Mean $= np = 12 \times \dfrac{4}{15} = 3.2$.

(c) Variance $= np(1 - p) = 12 \times \dfrac{4}{15} \times \dfrac{11}{15} = 2.347$.

Worked example 4.9

A group of 50 pensioners are all given a flu vaccination at their doctor's surgery. The probability that any one of this group will actually catch flu after this vaccination is known to be 0.1.

Find the distribution of X, the number of pensioners catching flu, and find the mean number and the variance of the number who catch flu.

Solution

The distribution is clearly binomial with $n = 50$ and $p = 0.1$.

So $X \sim \mathrm{B}(50, 0.1)$.

The mean number from the vaccinated group who will catch flu $= n \times p = 50 \times 0.1 = 5$.

The variance of the number catching flu $= n \times p \times (1 - p)$
$= 50 \times 0.1 \times 0.9 = 4.5$.

EXERCISE 4C

1 Components produced in a factory are tested in batches of 20. The proportion of components which are faulty is 0.2. Find the probability that a randomly chosen batch has:

(a) three or fewer faulty components,

(b) less than three faulty components,

(c) more than one faulty component.

2 A biased die, where the probability of a six showing is $\frac{2}{5}$, is thrown eight times. Find the probability that:

(a) a six shows fewer than three times,

(b) a six shows at least twice,

(c) no sixes show.

3 A group of 25 school pupils are asked to write an essay for a GCSE project. They each independently choose a subject at random from a selection of five. One of the choices available is to write a horror story. Find the probability that, out of this group,

(a) more than five write a horror story,

(b) at least six write a horror story,

(c) less than four write a horror story.

4 A cashier at a cinema has to calculate and balance the takings each evening. The probability that the cashier will make a mistake is 0.3. The manager of the cinema wishes to monitor the accuracy of the calculations over a 25-day working month. Find the probability that the cashier makes:

(a) fewer than five mistakes,

(b) no more than eight mistakes,

(c) more than three mistakes.

5 A manufacturer of wine glasses sells them in presentation boxes of 20. Random samples show that three in every hundred of these glasses are defective. Find the probability that a randomly chosen box contains:

(a) no defective glasses,

(b) at least two defective glasses,

(c) fewer than three defective glasses,

(d) exactly one defective glass.

6 The probability that a certain type of vacuum tube will shatter during a thermal shock test is 0.15. What is the probability that, if 25 such tubes are tested:

(a) four or more will shatter,

(b) no more than five will shatter,

(c) between five and ten (inclusive) will shatter?

7 A researcher calls at randomly chosen houses in a large city and asks the householder whether they will agree to answer questions on local services. The probability that a householder will refuse to answer the questions is 0.2. What is the probability that, on a day when 12 households are visited,

(a) three or fewer will refuse,

(b) exactly three will refuse,

(c) no more than one will refuse,

(d) at least ten **will agree** to answer?   [A]

8 The organiser of a school fair has organised raffle tickets to be offered to all adults who attend. The probability that an adult declines to buy a ticket is 0.15. What is the probability that if 40 adults attend and are asked to buy tickets:

(a) five or fewer will decline,

(b) exactly seven will decline,

(c) between four and ten (inclusive) decline,

(d) 36 or more **will agree** to buy.   [A]

9 A gardener plants beetroot seeds. The probability of a seed not germinating is 0.35, independently for each seed.

Find the probability that, in a row of 40 seeds, the number not germinating is:

(a) nine or fewer,

(b) seven or more,

(c) equal to the number germinating.   [A]

10 Safety inspectors carry out checks on the fire resistance of office furniture. Twenty per cent of items of furniture checked fail the test.

(a) Three randomly selected pieces of furniture are tested. Find the probability that:
(i) none of these three items will fail the test,
(ii) exactly one of these three items will fail the test.

(b) Twenty pieces of office furniture awaiting test include exactly four which will fail. If three of these twenty pieces, selected at random, are tested, find the probability that:
(i) all three will pass the test,
(ii) exactly two will pass the test.   [A]

4.7 The binomial model

In section 4.2, the essential elements of the binomial distribution are outlined.

Whilst it may be clear that a situation has a fixed number of trials and that two outcomes only are involved, often it can be a more difficult decision to make the assumption that the probability of each outcome is fixed and that the trials are independent of each other.

Examining situations to assess the suitability of the binomial model is an important aspect of this topic.

Worked example 4.10

A monkey in a cage is rewarded with food if it presses a button when a light flashes. Say, giving a reason, whether it is likely that the following variables follow the binomial distribution:

(a) X, the number of times that the light flashes before the monkey is twice successful in obtaining the food.

(b) Y, the number of times that the monkey obtains food by the time the light has flashed 20 times.

Solution

(a) In this case, the number of trials is clearly not fixed as we do not know how many times in total the light will need to flash in order for the monkey to achieve two successes. Therefore, it is immediately clear that a binomial model is not appropriate.

(b) Here, we clearly have a fixed value for n, the total number of trials involved. We have $n = 20$. However, we cannot assume that the probability of the monkey pressing the button to obtain food will remain constant. It would seem very likely that, once the monkey has made the connection between the light flashing, pressing the button and obtaining food, it will learn that a connection exists and the probability that it will press the button in subsequent trials in order to obtain food will increase. Therefore, as the probability is not constant, the binomial model is not appropriate.

Worked example 4.11

For each of the following experiments, state, giving reasons, whether a binomial distribution is appropriate.

Experiment 1

A bag contains black, white and red marbles which are selected at random, one at a time, with replacement. Ten marbles are taken out of the bag and the colour of each marble is noted.

Experiment 2

This experiment is a repeat of experiment 1 except that the bag contains black and white marbles only.

Experiment 3

This experiment is a repeat of experiment 2 except that the marbles are not replaced after selection.

4

Solution

In each experiment, the total number of trials is fixed at $n = 10$.

Experiment 1 cannot be modelled using the binomial distribution because there are **three outcomes** (black, white and red) being considered.

Experiment 2 has only two outcomes involved, black or white, and the probability of obtaining each outcome is fixed because the marbles are replaced after each trial. The trials are independent of each other so a binomial model would be suitable.

Experiment 3 cannot be modelled using the binomial distribution because the probability of each outcome is not constant since the marbles are not being replaced into the bag after each trial.

Worked example 4.12

A consultant wishes to test a new treatment for a particular skin condition. She asks patients who come to her clinic suffering from this condition if they are willing to take part in a trial in which they will be randomly allocated to the standard treatment or to the new treatment. She observes that the probability of such a patient agreeing to take part in this trial is 0.3.

(Assume throughout this question that the behaviour of any patient is independent of the behaviour of all other patients.)

(a) If she asks 25 patients to take part in the trial, find the probability that

  (i) four or fewer will agree,

  (ii) exactly six will agree,

  (iii) more than two will agree.

During one year, this doctor finds a total of 40 patients who are willing to join the trial.

There are many reasons why patients, who have agreed to take part in the trial, may withdraw before the end of the trial. It is found that the probability of a patient who receives the standard treatment withdrawing before the end of the trial is 0.14.

(b) Out of the 15 patients who are receiving the standard treatment, find the probability that no more than one withdraws before the end of the trial.

The probability that a patient who is receiving the new treatment withdraws before the end of the trial is 0.22.

(c) For each of the following cases, state, giving a reason, whether or not the binomial distribution is likely to provide an adequate model for the random variable R.

   (i) R is the total number of patients (out of 40) withdrawing before the end of the trial.

   (ii) R is the total number of patients asked in order to obtain the 40 to take part in the trial.

Solution

(a) In this case $X \sim B(25, 0.3)$ and cumulative binomial tables can be used.

   (i) $P(X \leq 4)$ is required which is obtained directly from the tables $= 0.0905$.

   (ii) $P(X = 6)$ can be evaluated directly as

$$\binom{25}{6} \times 0.3^6 \times 0.7^{19} = 0.1472 \text{ (four decimal places)}$$

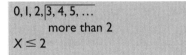

   or obtained from the tables by finding

$$P(X \leq 6) - P(X \leq 5) = 0.3407 - 0.1935$$
$$= 0.1472.$$

   (iii) $P(X > 2) = 1 - P(X \leq 2) \qquad = 1 - 0.009 \text{ (from tables)}$
   $$= 0.991.$$

(b) In this case $X \sim B(15, 0.14)$ and we require $P(X \leq 1)$.

   We do not have $n = 15$ and $p = 0.14$ available in the tables and so must evaluate $P(X = 0) + P(X = 1)$

$$= \binom{15}{0} \times 0.14^0 \times 0.86^{15} + \binom{15}{1} \times 0.14^1 \times 0.86^{14}$$

$$= 0.1041 + 0.2542$$

$$= 0.358$$

(c) (i) Binomial will not be a suitable model for R as p is not constant. The probability of a patient withdrawing is 0.14 if they received the standard treatment and 0.22 if they are on the new treatment.

(ii) Binomial will not be a suitable model for R as n is not fixed. We have no idea how many patients in total may need to be asked.

EXERCISE 4D

1 A tour operator organises a trip for cricket enthusiasts to the Caribbean in March. The package includes a ticket for a one-day International in Jamaica. Places on the tour must be booked in advance. From past experience, the tour operator knows that the probability of a person who has booked a place subsequently withdrawing is 0.08 and is independent of other withdrawals.

(a) Twenty people book places. Find the probability that:
  (i) none withdraw,
  (ii) two or more withdraw,
  (iii) exactly two withdraw.

(b) The tour operator accepts 22 bookings but has only 20 tickets available for the one-day International. What is the probability that he will be able to provide tickets for everyone who goes on tour? [A]

2 The organiser of a fund raising event for a sports club finds that the probability of a person who is asked to buy a raffle ticket refusing is 0.15.

(a) What is the probability that, if 40 people are asked to buy raffle tickets:
  (i) five or fewer will refuse,
  (ii) exactly seven will refuse,
  (iii) more than four will refuse.

The club also owns a fruit machine for the use of members. Inserting a 20p coin enables a member of the club to attempt to win a prize. The probability of winning a prize is a constant 0.2.

(b) Find the probability that a member, who has 25 attempts at winning on this machine, gains:
  (i) three or more prizes,
  (ii) no more than five prizes.

Another member of the club asks people to pay £1 to enter a game of chance. She continues to ask until 50 people have agreed to participate.

(c) X is the number of people she asks before obtaining the 50 participants. Say, giving a reason whether it is likely that X will follow a binomial distribution. [A]

3 An examination consists of 25 multiple choice questions each with five alternatives only one of which is correct and allocated one mark.

   (a) No marks are subtracted for incorrect answers. For a candidate who guesses the answers to all 25 questions, find:
   (i) the probability of obtaining at most eight marks,
   (ii) the probability of obtaining more than 12 marks,
   (iii) the probability of obtaining exactly ten marks.

   (b) It is decided to subtract one mark for each incorrect answer and to award four marks for each correct answer. Find the mean mark:
   (i) for a candidate who guesses all the answers,
   (ii) for a candidate whose chance of correctly answering each question is 0.8. [A]

4 A railway company employs a large number of drivers. During a dispute over safety procedures, the drivers consider taking strike action.

   Early in the dispute, a polling organisation asks a random sample of 20 of the drivers employed by the company whether they are in favour of strike action.

   (a) If the probability of a driver answering 'yes' is 0.4 and is independent of the answers of the other drivers, find the probability that ten or more drivers answer 'yes'.

   Later in the dispute, the probability of a driver answering 'yes' rises to 0.6.

   (b) If the polling organisation asks the same question to a second random sample of 20 drivers, find the probability that ten or more drivers answer 'yes'.

   A union meeting is now called and attended by 20 drivers. At the end of the meeting, those drivers in favour of strike action are asked to raise their hands.

   (c) Give **two** reasons why the probability distribution you used in **(b)** is unlikely to be suitable for determining the probability that ten or more of these 20 drivers raise their hands. [A]

5 Siballi is a student who travels to and from university by bus. He believes that the probability of having to wait more than six minutes to catch a bus is 0.4 and is independent of the time of day and direction of travel.

   (a) Assuming that Siballi's beliefs are correct, find the probability that, during a particular week when he catches ten buses, the number of times he has to wait more than six minutes is:

(i) three or fewer,

(ii) more than four.

(b) Assuming that Siballi's beliefs are correct, calculate values for the mean and standard deviation of the number of times he has to wait more than six minutes to catch a bus in a week when he catches ten buses.

(c) During a thirteen week period, the numbers of times (out of ten) he had to wait more than six minutes to catch a bus were as follows:

    4 8 8 9 3 2 2 7 0 1 5 2 0

(i) Calculate the mean and standard deviation of these data.

(ii) State, giving reasons, whether your answers to **(c)(i)** support Siballi's beliefs that the probability of having to wait more than six minutes to catch a bus is 0.4 and is independent of the time of day and direction of travel.   [A]

6 Dwight, a clerical worker, is employed by a benefits agency to calculate the weekly payments due to unemployed adults claiming benefit. These payments vary according to personal circumstances. During the first week of his employment, the probability that he calculates a payment incorrectly is 0.25 for each payment.

(a) Given that Dwight calculates 40 payments during his first week of employment, find the probability that:

(i) five or fewer are incorrect,

(ii) more than 30 are **correct**.

(b) The random variable, R, represents the number of payments he calculates until a total of ten have been carried out correctly. State, giving a reason, whether the binomial distribution is an appropriate model for R.

(c) A random sample of 40 payments is taken from those calculated by Dwight during his first **year** of employment. Give a reason why a binomial distribution may not provide a suitable model for the number of incorrect payments in the sample.

Key point summary

1 The binomial distribution can only be used to model   *p59*
situations in which certain conditions exist. These
conditions are:

- a fixed number of trials,
- two possible outcomes only at each trial,
- fixed probabilities for each outcome,
- trials independent of each other.

2 The parameters involved in describing a binomial   *p60*
distribution are **n** and **p**, where **n** represents the
number of trials and **p** represents the probability of
a success.

This is written $X \sim B(n, p)$.

3 The formula for evaluating a binomial probability   *p64*
of x successes out of n trials when the probability
of a success is p is

$$P(X = x) = \binom{n}{x} \times p^x \times (1 - p)^{(n - x)}$$

where $\binom{n}{x}$, which can also be written as ^nC_x is found

on most calculators.

It can also be found from Pascal's triangle or from
the definition

$$\frac{n!}{x!(n - x)!}.$$

4 Cumulative binomial tables can be used for   *p66*
evaluating binomial probabilities of the type
$P(X \le x)$ for certain values of n and p.

Use these tables if you can as it will save a lot
of time.

Remember to choose $p \le 0.5$

5 The mean and variance of the binomial   *pp68, 69*
distribution are given by

Mean $= np$

and

Variance $= np(1 - p)$.

Test yourself	What to review
1 Teenage girls are known to declare a fear of snakes with probability 0.4.	*Section 4.4*
If a group of four teenage girls are chosen at random from a large school, what is the probability that exactly two of them will be afraid of snakes?	
2 A recent survey suggested that the proportion of 14-year-old boys who never consider their health when deciding what to eat is 0.2. What is the probability that, in a random sample of 20 14-year-old boys, the number who never consider their health is	*Section 4.5*
(a) exactly three,	
(b) exactly two?	
3 Among the blood cells of a particular animal species, it is known that the proportion of cells of type α is 0.4.	*Section 4.5*
In a sample of 30 blood cells from this species, find the probability that:	
(a) ten or fewer are of type α,	
(b) at most eight are of type α,	
(c) less than 15 are of type α.	
4 A market research company carried out extensive research into whether people believed in astrological predictions. It was found that the probability a person selected at random did believe in such predictions was 0.3.	*Section 4.5*
Out of a group of 15 randomly selected people, what is the probability that:	
(a) more than four believed,	
(b) at least six believed,	
(c) two or more believed?	
5 A nationwide survey discovered that 3% of the population believed that infants should be fed only on organic produce.	*Section 4.7*
A group of 12 mothers, four of whom were selected from members at a health club and the remaining eight from customers at a large supermarket, were invited to a meeting and asked to raise their hand if they felt that infants should be fed only on organic produce. Give two reasons, why the number raising their hands should not be modelled using a binomial distribution.	

4

Test yourself (*continued*)	What to review

6 In a manufacturing process, it is observed that 8% of the rods produced do not conform to the required specifications.

A sample of 40 components is taken at random from a large batch. What is the mean number of components not conforming in this sample and also, what is the variance?

Section 4.6

7 A car repair firm has eight enquiries for estimates one day. Over many years, the manager has found that the probability that any estimate will be accepted is 0.15. Calculate the probability that, out of the eight estimates given:

(a) exactly two will be accepted,

(b) more than half will be refused.

Write down the mean and standard deviation of the number accepted.

Sections 4.5 and 4.6

Test yourself **ANSWERS**

1 $X \sim B(4, 0.4)$,

$$P(X = 2) = \binom{4}{2}(0.4)^2(0.6)^2 = 0.346.$$

2 $X \sim B(20, 0.2)$ use tables.

(a) $P(X = 3) = 0.2053$;

(b) $P(X = 2) = 0.1369$.

3 $X \sim B(30, 0.4)$ use tables.

(a) $P(X \leq 10) = 0.2915$;

(b) $P(X \leq 8) = 0.0940$;

(c) $P(X < 15) = P(X \leq 14) = 0.8246$.

4 $A \sim B(15, 0.3)$ use tables.

(a) $P(A > 4) = 1 - P(A \leq 4) = 1 - 0.5155 = 0.4845$;

(b) $P(A \geq 6) = 1 - P(A \leq 5) = 1 - 0.7216 = 0.2784$;

(c) $P(A \geq 2) = 1 - P(A \leq 1) = 1 - 0.0353 = 0.9647$.

5 The value of p will probably be high at the health club and lower at the supermarket. A show of hands would mean that the trials are not independent, as people will be influenced by the way other people voted.

6 Mean $= 40 \times 0.08 = 3.2$.

Variance $= 40 \times 0.08 \times 0.92 = 2.944$.

7 $X \sim B(8, 0.15)$.

(a) $P(X = 2) = P(X \leq 2) - P(X \leq 1) = 0.8948 - 0.6572 = 0.2376$;

(b) $P(X \leq 3) = 0.9786$.

Mean of $X = 8 \times 0.15 = 1.2$.

Standard deviation of $X = \sqrt{8 \times 0.15 \times 0.85} = 1.010$.

CHAPTER 5

The normal distribution

Learning objectives

After studying this chapter, you should be able to:
- use tables to find probabilities from any normal distribution
- use tables of percentage points of the normal distribution
- understand what is meant by the distribution of the sample mean
- find probabilities involving sample means.

5.1 Continuous distributions

In the previous chapter you have studied a discrete probability distribution where the possible outcomes can be listed and a probability associated with each. For continuous variables such as height, weight or distance it is not possible to list all the possible outcomes. In this case probability is represented by the area under a curve (called the probability density function).

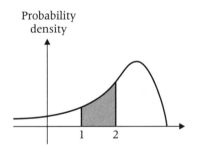

The probability that an observation, selected at random from the distribution, lies between 1 and 2 is represented by the shaded area. Note that the probability that an observation from a continuous distribution is exactly equal to 2 (or any other value) is zero.

There are two conditions for a curve to be used as a probability density function:

- the total area under the curve must be 1.

- the curve must not take negative values, that is, it must not go below the horizontal axis.

5.2 The normal distribution

Many continuous variables, which occur naturally, have a probability density function like this.

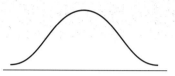

This is called a normal distribution. It has a high probability density close to the mean and this decreases as you move away from the mean.

> The main features of normal distribution are that it is:
> - bell shaped
> - symmetrical (about the mean)
> - the total area under the curve is 1 (as with all probability density functions).

Examples of variables which are likely to follow a normal distribution are the heights of adult females in the United Kingdom, the lengths of leaves from oak trees, the widths of car doors coming off a production line and the times taken by 12-year-old boys to run 100 m.

5.3 The standard normal distribution

Normal distributions may have any mean and any standard deviation. The normal distribution with mean 0 and standard deviation 1 is called the **standard normal distribution**.

> The equation of the probability density function (p.d.f.) is
> $$f(z) = \frac{1}{\sqrt{2\pi}} e^{-\frac{z^2}{2}} \quad \text{for } -\infty < z < \infty.$$

Z is, by convention, used to denote the standard normal variable.

Finding areas under this curve would involve some very difficult integration. Fortunately this has been done for you and the results tabulated. The tables are in the Appendix and an extract is shown in the next section.

AQA Formulae Book, Table 3.

5.4 The normal distribution function

The table gives the probability, p, that a normally distributed random variable Z, with mean = 0 and variance = 1, is less than or equal to z.

> z is used to denote a particular value of Z. You will not be penalised for failing to distinguish between z and Z.

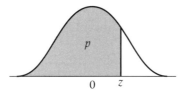

0.06
row ↓

z	0.00	0.01	0.02	0.03	0.04	0.05	0.06	0.07	0.08	0.09	
1.2	0.88493	0.88688	0.88877	0.89065	0.89251	0.89435	0.89617	0.89796	0.89973	0.90147	**1.2**
1.3	0.90320	0.90490	0.90658	0.90824	0.90988	0.91149	0.91309	0.91466	0.91621	0.91774	**1.3**
1.4	0.91924	0.92073	0.92220	0.92364	0.92507	0.92647	0.92785	0.92922	0.93056	0.93189	**1.4**
1.5	0.93319	0.93448	0.93574	0.93699	0.93822	0.93943	0.94062	0.94179	0.94295	0.94408	**1.5**
1.6	0.94520	0.94630	0.94738	0.94845	0.94950	0.95053	0.95154	0.95254	0.95352	0.95449	**1.6**

1.3 row →

As the diagram above shows, the area to the left of a particular value of z is tabulated. This represents the probability, p, that an observation, selected at random from a standard normal distribution (i.e. mean 0, standard deviation 1), will be less than z.

> The probability of an observation $< z$ is the same as the probability of an observation $\leqslant z$.

To use these tables for positive values of z, say **1.36**, take the digits before and after the decimal point and locate the appropriate row of the table – in this case the row, where z is **1.3** (see diagram in the margin). Then look along this row to find the probability in the column headed **0.06**. This gives 0.913 09 meaning that the probability that an observation from a standard normal distribution is less than 1.36 is 0.913 09.

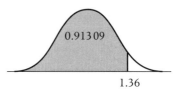

0.913 09

1.36

If the z-value is given to more than two decimal places, say 0.468, the appropriate value of p will lie between 0.677 24 (the value for $z = 0.46$) and 0.680 82 (the value for $z = 0.47$). An exact value could be estimated using interpolation. However, it is easier and perfectly acceptable to round the z to 0.47 and then use the tables.

> We will never need a final answer correct to five significant figures but if this is an intermediate stage of a calculation as many figures as possible should be kept.

EXERCISE 5A

Find the probability that an observation from a standard normal distribution will be less than:

(a) 1.23, **(b)** 0.97, **(c)** 1.85, **(d)** 0.42, **(e)** 0.09,

(f) 1.57, **(g)** 1.94, **(h)** 0.603, **(i)** 2.358, **(j)** 1.053 79.

Probability greater than *z*

The probability of a value greater than 1.36 is represented by the area to the right of 1.36. Use the fact that the total area under the curve is 1.

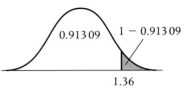

In this case $P(z > 1.36) = 1 - 0.913\,09 = 0.0869$.

EXERCISE 5B

1 Find the probability that an observation from a standard normal distribution will be greater than:

    (a) 1.36,   **(b)** 0.58,   **(c)** 1.23,   **(d)** 0.86,

    (e) 0.32,   **(f)** 1.94,   **(g)** 2.37,   **(h)** 0.652,

    (i) 0.087,   **(j)** 1.3486.

Negative values of *z*

Negative values of *z* are not included in the tables. This is because we can use the fact that the normal distribution is symmetrical to derive them from the positive values.

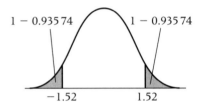

For example, $P(z < -1.52) = P(z > 1.52)$

    $P(z < -1.52) = 1 - 0.935\,74 = 0.064\,26$

Similarly, $P(z > -0.59) = P(z < 0.59) = 0.722\,40$.

EXERCISE 5C

1 Find the probability that an observation from a standard normal distribution will be:

    (a) less than -1.39,

    (b) less than -0.58,

    (c) more than -1.09,

    (d) more than -0.47,

    (e) less than or equal to -0.45,

    (f) greater than or equal to -0.32,

    (g) less than -0.64,

    (h) -0.851 or greater,

    (i) more than -0.747,

    (j) less than -0.4398.

5.5 Probability between z-values

To find the probability that z lies between two values we may have to use both symmetry and the fact that the total area under the curve is 1. It is essential to draw a diagram. Remember that the mean of a standard normal distribution is 0 and so positive values are to the right of the mode and negative values to the left. The following three examples cover the different possibilities.

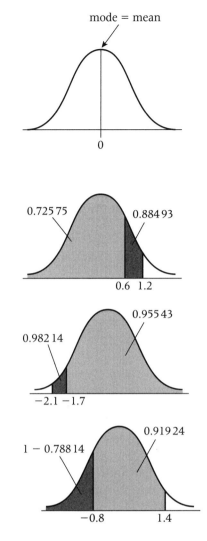

- $P(0.6 < z < 1.2)$

  We require:

     (area to left of 1.2) − (area to left of 0.6)

     $= 0.884\,93 - 0.725\,75$

     $= 0.159$

- $P(-2.1 < z < -1.7)$

  Here the z-values are negative, and so although we could still use areas to the left of z it is easier to use:

     (area right of −2.1) − (area right of −1.7)

     $= 0.982\,14 - 0.955\,43$

     $= 0.0267$

- $P(-0.8 < z < 1.4)$

  We require:

     (area left of 1.4) − (area left of −0.8)

     $= 0.919\,24 - (1 - 0.788\,14)$

     $= 0.707$

EXERCISE 5D

1 Find the probability that an observation from a standard normal distribution will be between:

  (a) 0.2 and 0.8,

  (b) −1.25 and −0.84,

  (c) −0.7 and 0.7,

  (d) −1.2 and 2.4,

  (e) 0.76 and 1.22,

  (f) −3 and −2,

  (g) −1.27 and 2.33,

  (h) 0.44 and 0.45,

  (i) −1.2379 and −0.8888,

  (j) −2.3476 and 1.9987.

5.6 Standardising a normal variable

The wingspans of a population of birds are normally distributed with mean 14.1 cm and standard deviation 1.7 cm. We may be asked to calculate the probability that a randomly selected bird has a wingspan less than 17.0 cm. Tables of the normal distribution with mean 14.1 and standard deviation 1.7 do not exist. However, we can use tables of the standard normal distribution by first standardising the value of interest. That is, we express it as standard deviations from the mean.

For example, for a normal distribution with mean 50 cm and standard deviation 5 cm, a value of 60 cm is:

$$60 - 50 = 10 \text{ cm from the mean.}$$

To express this as standard deviations from the mean we divide by 5 cm.

$$\frac{10}{5} = 2$$

This is the standardised or z-score of 60 cm.

> For a value, x, from a normal distribution with mean μ and standard deviation σ,
>
> $$z = \frac{x - \mu}{\sigma}$$

For the distribution with mean 50 cm, standard deviation 5 cm, the z-score of 47 cm is $\frac{(47 - 50)}{5} = -0.6$. Note the importance

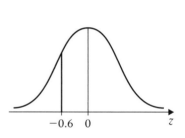

of the sign which tells us whether the value is to the left or the right of the mean.

EXERCISE 5E

1 A normal distribution has mean 40 cm and standard deviation 5 cm. Find the standardised values of:

 (a) 47 cm, **(b)** 43 cm, **(c)** 36 cm, **(d)** 32 cm,

 (e) 50.5 cm.

2 A normal distribution has mean 36.3 s and standard deviation 4.6 s. Find the z-scores of:

 (a) 39.3 s, **(b)** 30.0 s, **(c)** 42.5 s, **(d)** 28.0 s.

3 The wingspans of a population of birds are approximately normally distributed with mean 18.1 cm and standard deviation 1.8 cm. Find standardised values of:

 (a) 20.2 cm, **(b)** 17.8 cm, **(c)** 19.3 cm, **(d)** 16.0 cm.

5.7 Probabilities from a normal distribution

To find the probability that a bird randomly selected from a population with mean wingspan 14.1 cm and standard deviation 1.7 cm, has a wingspan less than 17 cm, first calculate the z-score:

$$z = \frac{(17 - 14.1)}{1.7} = 1.71$$

Now enter the tables at 1.71.

We find that the probability of a wingspan less than 17 cm is 0.956.

Some students wonder whether less than 17 cm really means less than 16.5 cm. **Don't.** Just use the value given. Otherwise you would also have to say that the standard deviation is between 1.65 and 1.75 cm and the calculation becomes impossible to carry out.

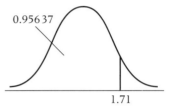

Worked example 5.1

The chest measurements of adult male customers for T-shirts may be modelled by a normal distribution with mean 101 cm and standard deviation 5 cm. Find the probability that a randomly selected customer will have a chest measurement which is:

(a) less than 103 cm,

(b) 98 cm or more,

(c) between 95 cm and 100 cm,

(d) between 90 cm and 110 cm.

Some calculators will find these probabilities directly. However, no correctly available calculator will answer questions such as worked examples 5.3, 5.4, and 5.5. It is still necessary to know how to standardise variables and to use tables.

5

Solution

(a) $z = \frac{(103 - 101)}{5} = 0.4$,

probability less than 103 cm is 0.655;

(b) $z = \frac{(98 - 101)}{5} = -0.6$,

probability 98 cm or more is 0.726;

(c) $z_1 = \frac{(95 - 101)}{5} = -1.2$,

$z_2 = \frac{(100 - 101)}{5} = -0.2$,

probability between 95 cm and 100 cm is $0.884\,93 - 0.579\,26 = 0.306$;

(d) $z_1 = \frac{(90 - 101)}{5} = -2.2$,

$z_2 = \frac{(110 - 101)}{5} = 1.8$,

probability between 90 cm and 101 cm is $0.964\,07 - (1 - 0.986\,10) = 0.950$.

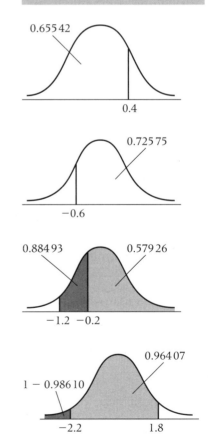

EXERCISE 5F

(In this exercise give the answers correct to three significant figures or to the accuracy found from tables if this is less than three significant figures.)

1 A variable is normally distributed with a mean of 19.6 cm and a standard deviation of 1.9 cm. Find the probability that an item chosen at random from this distribution will have a measurement:

(a) less than 20.4 cm,

(b) more than 22.0 cm,

(c) 17.5 cm or less,

(d) 22.6 cm or less,

(e) between 19.0 and 21.0 cm,

(f) between 20.5 cm and 22.5 cm.

2 The weights of a certain animal are approximately normally distributed with a mean of 36.4 kg and a standard deviation of 4.7 kg. Find the probability that when one of these animals is chosen at random it will have a weight:

(a) 40.0 kg or less,

(b) between 32.0 kg and 41.0 kg,

(c) more than 45.0 kg,

(d) less than 28.0 kg,

(e) 30.0 kg or more,

(f) between 30.0 kg and 35.0 kg.

3 The weights of the contents of jars of jam packed by a machine are approximately normally distributed with a mean of 460.0 g and a standard deviation of 14.5 g. A jar of jam is selected at random. Find the probability that its contents will weigh:

(a) less than 450 g,

(b) 470.0 g or less,

(c) between 440.0 and 480.0 g,

(d) 475 g or more,

(e) more than 454 g,

(f) between 450 g and 475 g.

4 The lengths of leaves from a particular plant are approximately normally distributed with a mean of 28.4 cm and a standard deviation of 2.6 cm. When a leaf is chosen at random what is the probability its length is:

(a) between 25.0 cm and 30.0 cm,

(b) more than 32.0 cm,

(c) less than 24.0 cm,

(d) 27.0 cm or more,

(e) 26.0 cm or less,

(f) between 24.0 cm and 28.0 cm?

5 The volumes of the discharges made by a drink dispensing machine into cups is approximately normally distributed with a mean of 465.0 cm^3 and a standard deviation of 6.8 cm^3. When the volume of the contents of a cup chosen at random from this machine is measured what is the probability that it will be:

(a) 470 cm^3 or more,

(b) less than 458.0 cm^3,

(c) between 455.0 cm^3 and 475.0 cm^3?

5.8 Percentage points of the normal distribution

This is an alternative way of tabulating the standard normal distribution. The z-score for a given probability, p, is tabulated.

The table gives the values of z satisfying $P(Z \leq z) = p$, where Z is the normally distributed random variable with mean $= 0$ and variance $= 1$.

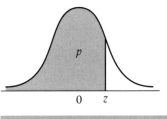

AQA Formulae Book, Table 4.

0.00 column

p	0.00	0.01	0.02	0.03	0.04	0.05	0.06	0.07	0.08	0.09	
0.5	0.0000	0.0251	0.0502	0.0753	0.1004	0.1257	0.1510	0.1764	0.2019	0.2275	**0.5**
0.6	0.2533	0.2793	0.3055	0.3319	0.3585	0.3853	0.4125	0.4399	0.4677	0.4958	**0.6**
0.7	0.5244	0.5534	0.5828	0.6128	0.6433	0.6745	0.7063	0.7388	0.7722	0.8064	**0.7**
0.8	0.8416	0.8779	0.9154	0.9542	0.9945	1.0364	1.0803	1.1264	1.1750	1.2265	**0.8**
0.9	1.2816	1.3408	1.4051	1.4758	1.5548	1.6449	1.7507	1.8808	2.0537	2.3263	**0.9**

0.9 row →

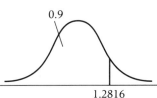

If we require the z-score which exceeds 0.9, or 90%, of the normal distribution, we would locate the row **0.9** and then take the entry in the column **0.00**. This gives a z-score of 1.2816.

Values of p less than 0.5 are not tabulated. To find the z-score which exceeds 0.05, or 5%, of the distribution, we need to use symmetry. The z-score will clearly be negative but will be of the same magnitude as the z-score which exceeds $1 - 0.05 = 0.95$ of the distribution. Thus, the required value is -1.6449.

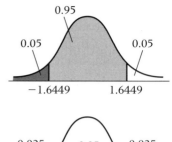

We often require to find the z-scores which are symmetrical about the mean and contain 95% of the distribution. The two tails will contain 5% in total. They will therefore contain $\frac{5}{2} = 2.5\%$ each. The upper z-score will exceed $100 - 2.5 = 97.5\%$ of the distribution. Entering the table at 0.975 we find a z-score of 1.96. The lower z-score is, by symmetry, -1.96.

p	0.00	0.01	0.02	0.03	0.04	0.05	0.06	0.07	0.08	0.09	
0.95	1.6449	1.6546	1.6646	1.6747	1.6849	1.6954	1.7060	1.7169	1.7279	1.7392	0.95
0.96	1.7507	1.7624	1.7744	1.7866	1.7991	1.8119	1.8250	1.8384	1.8522	1.8663	0.96
0.97	1.8808	1.8957	1.9110	1.9268	1.9431	1.9600	1.9774	1.9954	2.0141	2.0335	0.97
0.98	2.0537	2.0749	2.0969	2.1201	2.1444	2.1701	2.1973	2.2252	2.2571	2.2094	0.98
0.99	2.3263	2.3656	2.4089	2.4573	2.5121	2.5758	2.6521	2.7478	2.8782	3.0902	0.99

EXERCISE 5G

1 Find the z-score which:

(a) is greater than 97.5% of the population,

(b) is less than 90% of the population,

(c) exceeds 5% of the population,

(d) is exceeded by 7.5% of the population,

(e) is greater than 2.5% of the distribution,

(f) is less than 15% of the population,

(g) exceeds 20% of the distribution,

(h) is greater than 90% of the distribution,

(i) is less than 1% of the population.

2 Find the z-scores which are symmetrical about the mean and contain:

(a) 90% of the distribution,

(b) 99% of the distribution,

(c) 99.8% of the distribution.

Applying results to normal distributions

To apply these results to normal distributions, other than the standard normal, we need to recall that z-scores are in units of standard deviations from the mean. Thus if x is normally distributed with mean μ and standard deviation σ,

$$x = \mu + z\sigma.$$

> Note that this is only a rearrangement of the formula $z = (x - \mu)/\sigma$.

Worked example 5.2

The wingspans of a population of birds are normally distributed with mean 14.1 cm and standard deviation 1.7 cm. Find:

(a) the wingspan which will exceed 90% of the population,

(b) the wingspan which will exceed 20% of the population,

(c) the limits of the central 95% of the wingspans.

Solution

(a) The z-score which exceeds 90% of the population is 1.2816. The value required is therefore 1.2816 standard deviations above the mean, i.e.

$$14.1 + 1.2816 \times 1.7 = 16.3 \text{ cm.}$$

(b) The z-score which will exceed 20% of the population will be exceeded by 80% of the population.

$$z = -0.8416$$
$$x = 14.1 - 0.8416 \times 1.7 = 12.7 \text{ cm.}$$

(c) $z = \pm 1.96$

The central 95% of wingspans are $14.1 \pm 1.96 \times 1.7$, i.e.

$$14.1 \pm 3.33 \text{ or } 10.8 \text{ cm to } 17.4 \text{ cm.}$$

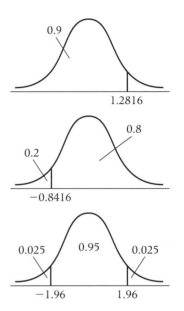

EXERCISE 5H

1 A large shoal of fish have lengths which are normally distributed with mean 74 cm and standard deviation 9 cm.

(a) What length will be exceeded by 10% of the shoal?

(b) What length will be exceeded by 25% of the shoal?

(c) What length will be exceeded by 70% of the shoal?

(d) What length will exceed 95% of the shoal?

(e) Find the limits of the central 90% of lengths.

(f) Find the limits of the central 60% of lengths.

2 Hamburger meat is sold in 1 kg packages. The fat content of the packages is found to be normally distributed with mean 355 g and standard deviation 40 g.

Find the fat content which will be exceeded by:

(a) 5% of the packages,

(b) 35% of the packages,

(c) 50% of the packages,

(d) 80% of the packages,

(e) 99.9% of the packages.

(f) Find the limits of the central 95% of the contents.

3 When Kate telephones for a taxi, the waiting time is normally distributed with mean 18 minutes and standard deviation 5 minutes. At what time should she telephone for a taxi if she wishes to have a probability of:

(a) 0.9 that it will arrive before 3.00 p.m.,

(b) 0.99 that it will arrive before 3.00 p.m.,

(c) 0.999 that it will arrive before 3.00 p.m.,

(d) 0.2 that it will arrive before 3.00 p.m.,

(e) 0.3 that it will arrive after 3.00 p.m.,

(f) 0.8 that it will arrive after 3.00 p.m.?

Worked example 5.3

A vending machine discharges soft drinks. A total of 5% of the discharges have a volume of more than 475 cm³, while 1% have a volume less than 460 cm³. The discharges may be assumed to be normally distributed. Find the mean and standard deviation of the discharges.

Solution

5% of z-scores exceed 1.6449.

Hence, if the mean is μ and the standard deviation is σ,

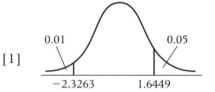

$$\mu + 1.6449\sigma = 475. \tag{1}$$

1% of z-scores are below -2.3263.

Hence,

$$\mu - 2.3263\sigma = 460. \tag{2}$$

Subtracting equation [2] from equation [1] gives

$$3.9712\sigma = 15$$
$$\sigma = 3.7772$$

Substituting in equation [1]

$$\mu + 1.6449 \times 3.7772 = 475$$
$$\mu = 468.787$$

The mean is 468.79 cm³ and the standard deviation is 3.78 cm³.

Worked example 5.4

Adult male customers for T-shirts have chest measurements which may be modelled by a normal distribution with mean 101 cm and standard deviation 5 cm. T-shirts to fit customers with chest measurements less than 98 cm are classified **small**. Find the median chest measurement of customers requiring **small** T-shirts.

Solution

First find the proportion of customers requiring **small** T-shirts.
$$z = \frac{(98 - 101)}{5} = -0.6$$

Proportion is $1 - 0.725\,75 = 0.274\,25$.

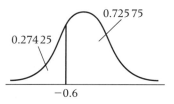

The chest measurement will be less than the median for half of these customers. That is, for,
$$\frac{0.274\,25}{2} = 0.137\,125 \text{ of all customers.}$$

The proportion of customers with chest measurements exceeding the median of those requiring small T-shirts is $1 - 0.137\,125 = 0.862\,875$.

The z-score is -1.08.

Thus, the median is $101 - 1.08 \times 5 = 95.6$ cm.

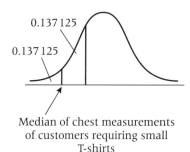

Median of chest measurements of customers requiring small T-shirts

5.9 Modelling data using the normal distribution

Textbooks and examination questions often use phrases such as 'the weights of packs of butter in a supermarket may be **modelled** by a normal distribution with mean 227 g and standard deviation 7.5 g'.

The word **modelled** implies that the weights may not follow a normal distribution exactly but that calculations which assume a normal distribution will give answers which are very close to reality. For example, if you use the normal distribution to calculate the proportion of packs which weigh less than 224 g, the answer you obtain will be very close to the proportion of packs which actually weigh less than 224 g.

There are at least two reasons why the word **modelled** is used in this context:

- We could never obtain sufficient data to prove that the weights followed a particular distribution exactly without the smallest deviation in any respect.

- The theoretical normal distribution does not have any limits. That is, it would have to be theoretically possible for the packs of butter to have any weight, including negative weights, to fit a normal distribution exactly. However, this is not a practical problem since, for a normal distribution with mean μ and standard deviation σ, the central:

     68% of the area lies in the range $\mu \pm \sigma$

     95.5% of the area lies in the range $\mu \pm 2\sigma$

     99.7% of the area lies in the range $\mu \pm 3\sigma$

For example, for the packs of butter we would expect 99.7% to lie in the range

$$227 \pm 3 \times 7.5 \quad \text{i.e. } 204.5\,\text{g to } 249.5\,\text{g.}$$

It would be theoretically possible to find a pack weighing 260 g which is well outside this range. However, this is so unlikely that, if we did, it would be sensible to conclude that the model was incorrect.

EXERCISE 5I

1 Shoe shop staff routinely measure the length of their customers' feet. Measurements of the length of one foot (without shoes) from each of 180 adult male customers yielded a mean length of 29.2 cm and a standard deviation of 1.47 cm.

   Given that the lengths of male feet may be modelled by a normal distribution, and making any other necessary assumptions, calculate an interval within which 90% of the lengths of male feet will lie.

2 Consultants employed by a large library reported that the time spent in the library by a user could be modelled by a normal distribution with mean 65 minutes and standard deviation 20 minutes.

   (a) Assuming that this model is adequate, what is the probability that a user spends:
   (i) less than 90 minutes in the library,
   (ii) between 60 and 90 minutes in the library?

   The library closes at 9.00 p.m.

   (b) Explain why the model above could not apply to a user who entered the library at 8.00 p.m.

   (c) Estimate an approximate latest time of entry for which the model above could still be plausible. [A]

3 The bar receipts at a rugby club after a home league game may be modelled by a normal distribution with mean £1250 and standard deviation £210.

   The club treasurer has to pay a brewery account of £1300 the day after the match.

   (a) What is the probability that she will be able to pay the whole of the account from the bar receipts?

Instead of paying the whole of the account she agrees to pay the brewery £x.

(b) What value of x would give a probability of 0.99 that the amount could be met from the bar receipts?

(c) What is the probability that the bar receipts after four home league games will all exceed £1300?

(d) Although the normal distribution may provide an adequate model for the bar receipts, give a reason why it cannot provide an exact model.   [A]

5.10 Notation

Many textbooks use the notation $X \sim N(\mu, \sigma^2)$ to mean that the variable X is normally distributed with mean μ and standard deviation σ. The symbol σ^2 is the square of the standard deviation and, as we have seen earlier, this is called the variance.

The variance is not a natural measure of spread as it is in different units from the raw data. It does, however, have many uses in mathematical statistics. We will not use it further in this book but it does appear in later modules.

$X \sim N(27.0, 16.0)$ means that the variable X is distributed with mean 27.0 and standard deviation $\sqrt{16.0} = 4.0$.

5.11 The central limit theorem

A bakery makes loaves of bread with a mean weight of 900 g and a standard deviation of 20 g. An inspector selected four loaves at random and weighed them. It is unlikely that the mean weight of the four loaves she chose would be exactly 900 g. In fact the mean weight was 906 g. A second inspector then chose four loaves at random and found their mean weight to be 893 g. There is no limit to how many times a sample of four can be chosen and the mean weight calculated. These means will vary and will have a distribution.

> This distribution is known as **the distribution of the sample mean**.

This is one of the most important statistical ideas in this book. You may not find it easy to grasp at first but you will meet it in many different contexts and this will help you to understand it.

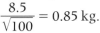 If a random sample of size n is taken from any distribution with mean μ and standard deviation σ then:

- $\bar{x}$, the sample mean, will be distributed with mean μ and standard deviation $\dfrac{\sigma}{\sqrt{n}}$,

- the distribution will be approximately normal provided n is sufficiently large – the larger the size of n the better the approximation.

> This result is exact. There is no approximation.

The second part of this result is known as the **central limit theorem**. It enables us to make statements about sample means without knowing the shape of the distribution they have come from. As a rule of thumb most textbooks say that the sample size, n, needs to be at least 30 to assume that the mean is normally distributed. For the purpose of examination questions it is best to stick to this figure, however it is undoubtedly on the cautious side. How large the sample needs to be depends on how much the distribution varies from the normal. For a *unimodal* distribution which is somewhat skew even samples of five or six will give a good approximation.

> If the parent distribution is normal, the distribution of the sample mean is exactly normal.

If a random sample of size 50 is taken from any distribution with mean 75.2 kg and standard deviation 8.5 kg then the mean will be approximately normally distributed with mean 75.2 kg and standard deviation $\dfrac{8.5}{\sqrt{50}} = 1.20$ kg.

If the sample is of size 100 the mean will be approximately normally distributed with mean 75.2 kg and standard deviation $\dfrac{8.5}{\sqrt{100}} = 0.85$ kg.

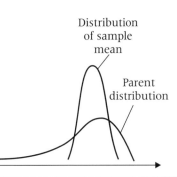

Distribution of sample mean

Parent distribution

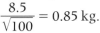 As the sample gets larger the standard deviation of the sample mean (sometimes called the **standard error**) gets smaller.

The sample means will be packed tightly around the population mean. The larger the samples become the tighter the means will be packed. This has major implications for topics which arise later, including confidence intervals, hypothesis testing and quality control.

Worked example 5.5

The weights of pebbles on a beach are distributed with mean 48.6 g and standard deviation 8.5 g.

(a) A random sample of 50 pebbles is chosen. Find the probability that:

    (i) the mean weight will be less than 49.0 g,

    (ii) the mean weight will be 47.0 g or less.

(b) Find limits within which the central 95% of such sample means would lie.

(c) How large a sample would be needed in order that the central 95% of sample means would lie in an interval of width at most 4 g?

Solution

(a) The distribution of the pebble weights is unknown but since the samples are of size 50 it is safe to use the central limit theorem and assume that the sample means are approximately normally distributed. This distribution of sample means will have mean 48.6 g and standard deviation $\frac{8.5}{\sqrt{50}} = 1.2021$ g.

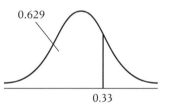

    (i) We first standardise 49.0

$$z = \frac{(49.0 - 48.6)}{(8.5/\sqrt{50})} = 0.333$$

    Note for the mean of samples of size n,

$$z = \frac{(\bar{x} - \mu)}{\left(\dfrac{\sigma}{\sqrt{n}}\right)}$$

> Note that we have rounded 0.333 to 0.33. This is adequate but not exact. A more accurate result could be found using interpolation.

    The probability that the mean is less than 49.0 g is 0.629.

    (ii) $z = \dfrac{(47.0 - 48.6)}{\left(\dfrac{8.5}{\sqrt{50}}\right)} = -1.331$

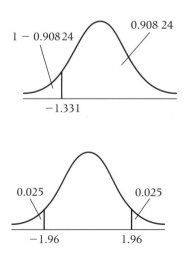

    The probability that the mean will be less than 47.0 g is $1 - 0.908\,24 = 0.0918$.

(b) The central 95% of sample means will lie in the

    interval $\mu \pm \dfrac{1.96\sigma}{\sqrt{n}}$,

    i.e. $48.6 \pm 1.96 \times \dfrac{8.5}{\sqrt{50}}$,

    or 46.2 g to 51.0 g.

(c) As in **(b)** the central 95% of sample means will lie in the interval $\mu \pm 1.96 \dfrac{\sigma}{\sqrt{n}}$. The width of this interval is

$$\frac{3.92 \times 8.5}{\sqrt{n}}, \quad \text{i.e.} \ \frac{33.32}{\sqrt{n}}.$$

If the interval is to be at most 4 then,

$$\frac{33.32}{\sqrt{n}} < 4$$

$$\text{i.e.} \ \frac{33.32}{4} < \sqrt{n}$$

$$69.4 < n$$

Thus a sample of size at least 70 is needed. (Fortunately the sample size has turned out to be quite large thus justifying our earlier assumption of a normally distributed sample mean.)

EXERCISE 5J

1 A population has a mean of 57.4 kg and a standard deviation of 6.7 kg. Samples of 80 items are chosen at random from this population. Find the probability that a sample mean:

(a) will be 58.4 kg or less,

(b) will be less than 56.3 kg,

(c) will lie between 56.3 kg and 58.4 kg.

2 It is found that the mean of a population is 46.2 cm and its standard deviation is 2.3 cm. Samples of 100 items are chosen at random.

(a) Between what limits would you expect the central 95% of the means from such samples to lie?

(b) What limit would you expect to be exceeded by only 5% of the sample means?

(c) How large should the sample size be in order for the central 95% of such sample means to lie in an interval of width at most 0.8 cm?

3 The times taken by people to complete a task are distributed with a mean of 18.0 s and a standard deviation 8.5 s. Samples of 50 times are chosen at random from this population.

(a) What is the probability that a randomly selected sample mean will:

(i) be at least 19.4 s,

(ii) be 17.5 s or more,

(iii) lie between 17.4 and 19.0 s?

(b) Between what limits would you expect the central 95% of such sample means to lie?

4 A population has a mean of 124.3 cm and a standard deviation of 14.5 cm.

   (a) What size samples should be chosen in order to make the central 95% of their means lie in an interval of width as close to 5.0 cm as possible?

   (b) Explain why, provided the samples are chosen at random, your answer is valid.

   (c) Under what circumstances might an answer to a similar question be invalid?

MIXED EXERCISE

1 A smoker's blood nicotine level, measure in ng/ml, may be modelled by a normal random variable with mean 310 and standard deviation 110.

   (a) What proportion of smokers have blood nicotine levels lower than 250?

   (b) What blood nicotine level is exceeded by 20% of smokers?   [A]

2 The lengths of components from a machine may be modelled by a normal distribution with mean 65 mm and standard deviation 2 mm. Find the probability that the length of a component selected at random will be less than 67 mm.   [A]

3 A health food cooperative markets free-range eggs. Eggs weighing less than 48 g are graded small, those weighing more than 59 g are graded large and the rest are graded medium.

   The weight of an egg from a particular supplier is normally distributed with mean 52 g and standard deviation 4 g.

   Find:

   (a) the proportion of eggs graded small,

   (b) the proportion of eggs graded medium,

   (c) the median weight of the eggs graded large.   [A]

4 Shamim drives from her home in Sale to college in Manchester every weekday during term. On the way she collects her friend David who waits for her at the end of his road in Chorlton. Shamim leaves home at 8.00 a.m., and the time it takes her to reach the end of David's road is normally distributed with mean 23 minutes and standard deviation 5 minutes.

   (a) Find the probability that she arrives at the end of David's road before 8.30 a.m.

   (b) If David arrives at the end of his road at 8.05 a.m. what is the probability that he will have to wait less than 15 minutes for Shamim to arrive?

5

(c) What is the latest time, to the nearest minute, that David can arrive at the end of his road to have a probability of at least 0.99 of arriving before Shamim? [A]

5 Free-range eggs supplied by a health food cooperative have a mean weight of 52 g with a standard deviation of 4 g. Assuming the weights are normally distributed find the probability that:

(a) a randomly selected egg will weigh more than 60 g,

(b) the mean weight of five randomly selected eggs will be between 50 g and 55 g,

(c) the mean weight of 90 randomly selected eggs will be between 52.1 g and 52.2 g.

Which of your answers would be unchanged if the weights are not normally distributed?

6 Bags of sugar are sold as 1 kg. To ensure bags are not sold underweight the machine is set to put a mean weight of 1004 g in each bag. The manufacturer claims that the process works to a standard deviation of 2.4 g. What proportion of bags are underweight?

7 The lengths of components produced by a machine are normally distributed with a mean of 0.984 cm and a standard deviation of 0.006 cm. The specification requires that the length should measure between 0.975 cm and 0.996 cm. Find the probability that a randomly selected component will meet the specification. [A]

8 The weights of bags of fertiliser may be modelled by a normal distribution with mean 12.1 kg and standard deviation 0.4 kg. Find the probability that:

(a) a randomly selected bag will weigh less than 12.0 kg,

(b) the mean weight of four bags selected at random will weigh more than 12.0 kg,

(c) the mean weight of 100 bags will be between 12.0 and 12.1 kg.

How would your answer to (c) be affected if the normal distribution was not a good model for the weights of the bags?

9 The weights of plums from an orchard have mean 24 g and standard deviation 5 g. The plums are graded small, medium or large. All plums over 28 g are regarded as large and the rest equally divided between small and medium. Assuming a normal distribution find:

(a) the proportion of plums graded large,

(b) the upper limit of the weights of the plums in the small grade. [A]

10 A survey showed that the value of the change carried by an adult male shopper may be modelled by a normal distribution with mean £3.10 and standard deviation £0.90. Find the probability that:

(a) an adult male shopper selected at random will be carrying between £3 and £4 in change,

(b) the mean amount of change carried by a random sample of nine adult male shoppers will be between £3.00 and £3.05.

Give two reasons why, although the normal distribution may provide an adequate model, it cannot in these circumstances provide an exact model.

11 The weights of pieces of home-made fudge are normally distributed with mean 34 g and standard deviation 5 g.

(a) What is the probability that a piece selected at random weighs more than 40 g?

(b) For some purposes it is necessary to grade the pieces as small, medium or large. It is decided to grade all pieces weighing over 40 g as large and to grade the heavier half of the remainder as medium. The rest will be graded as small. What is the upper limit of the small grade? [A]

12 A gas supplier maintains a team of engineers who are available to deal with leaks reported by customers. Most reported leaks can be dealt with quickly but some require a long time. The time (excluding travelling time) taken to deal with reported leaks is found to have a mean of 65 minutes and a standard deviation of 60 minutes.

(a) Assuming that the times may be modelled by a normal distribution, estimate the probability that:
  (i) it will take more than 185 minutes to deal with a reported leak,
  (ii) it will take between 50 minutes and 125 minutes to deal with a reported leak,
  (iii) the mean time to deal with a random sample of 90 reported leaks is less than 70 minutes.

(b) A statistician, consulted by the gas supplier, stated that, as the times had a mean of 65 minutes and a standard deviation of 60 minutes, the normal distribution would not provide an adequate model.
  (i) Explain the reason for the statistician's statement.
  (ii) Give a reason why, despite the statistician's statement, your answer to **(a)(iii)** is still valid.
[A]

13 A hot drinks machine delivers hot water into a cup when a button is pressed. The volume delivered may be modelled by a normal distribution with mean 470 ml and standard deviation 25 ml.

(a) Find the probability that the volume of hot water delivered will be:
(i) less than 480 ml,
(ii) between 450 ml and 480 ml.

(b) The volume of hot water delivered on a random sample of ten occasions is measured. Find the probability that the sample mean is less than 475 ml.

Each cup receiving the hot water has a capacity of 500 ml.

(c) Find the probability that when the button is pressed the volume of hot water delivered will exceed the capacity of the cup and so overflow.

It is possible to reset the machine so that the mean volume delivered takes any required value. The standard deviation remains unchanged.

(d) Find the mean value for the volume of hot water delivered in order that the probability of the cup overflowing is 0.001.

A new machine is bought. It is observed that for this machine the mean volume delivered is 490 ml and cups overflow on 15% of occasions.

(e) Assuming that the volume delivered by this machine may also be modelled by a normal distribution, find its standard deviation.

14 Yuk Ping belongs to an athletics club. In javelin throwing competitions her throws are normally distributed with mean 41.0 m and standard deviation 2.0 m.

(a) What is the probability of her throwing between 40 m and 46 m?

(b) What distance will be exceeded by 60% of her throws?

Gwen belongs to the same club. In competitions, 85% of her javelin throws exceed 35 m and 70% exceed 37.5 m. Her throws are normally distributed.

(c) Find the mean and standard deviation of Gwen's throws, each correct to two significant figures.

(d) The club has to choose one of these two athletes to enter a major competition. In order to qualify for the final round it is necessary to achieve a throw of at least 48 m in the preliminary rounds. Which athlete should be chosen and why? [A]

15 A machine is used to fill tubes, of nominal content 100 ml, with toothpaste. The amount of toothpaste delivered by the machine is normally distributed and may be set to any required mean value. Immediately after the machine has been overhauled, the standard deviation of the amount delivered is 2 ml. As time passes, this standard deviation increases until the machine is again overhauled.

The following three conditions are necessary for a batch of tubes of toothpaste to comply with current legislation:

(I) The average content of the tubes must be at least 100 ml.

(II) Not more than 2.5% of the tubes may contain less than 95.5 ml.

(III) Not more than 0.1% of the tubes may contain less than 91 ml.

(a) For a batch of tubes with mean content 98.8 ml and standard deviation 2 ml, find the proportion of tubes which contain:

    (i) less than 95.5 ml,

    (ii) less than 91 ml.

    Hence state which, if any, of the three conditions above are **not** satisfied.

(b) If the standard deviation is 5 ml, find the mean in **each** of the following cases:

    (i) exactly 2.5% of tubes contain less than 95.5 ml,

    (ii) exactly 0.1% of tubes contain less than 91 ml.

    Hence state the smallest value of the mean which would enable all three conditions to be met when the standard deviation is 5 ml.

(c) Currently exactly 0.1% of tubes contain less than 91 ml and exactly 2.5% contain less than 95.5 ml.

    (i) Find the current values of the mean and the standard deviation.

    (ii) State, giving a reason, whether you would recommend that the machine is overhauled immediately. [A]

5

Key point summary

I The normal distribution is continuous, symmetrical and bell-shaped.   *p82*

2 The normal distribution with mean 0 and standard deviation 1 is called the standard normal distribution. Tables of this distribution are in the Appendix.   *p82, 83*

3 An observation, x, from a normal distribution with mean μ and standard deviation σ is standardised using the formula   *p86*

$$z = \frac{x - \mu}{\sigma}.$$

This must be done before the tables can be used.

4 If a random sample of size n is taken from any distribution with mean μ and standard deviation σ, then: *p95*

- $\bar{x}$, the sample mean, will be distributed with mean μ and standard deviation $\frac{\sigma}{\sqrt{n}}$,

- the distribution will be approximately normal provided n is reasonably large.

Test yourself	What to review
1 Why do tables of the standard normal distribution not tabulate negative values of z?	*Sections 5.1 and 5.4*
2 For a standard normal distribution find the value of z which is exceeded with probability: **(a)** 0.06, **(b)** 0.92.	*Section 5.8*
3 A normal distribution has mean 12 and standard deviation 4. Find the probability that an observation from this distribution: **(a)** exceeds 10, **(b)** is less than 5, **(c)** is between 14 and 16, **(d)** is between 8 and 15.	*Section 5.7*
4 What is the probability that an observation from the distribution in question **3** is exactly equal to 10?	*Section 5.1*
5 Under what circumstances may tables of the normal distribution be useful when dealing with a variable which is not normally distributed?	*Section 5.11*
6 A random sample of size 25 is taken from a normal distribution with mean 20 and standard deviation 10. **(a)** Find the probability that the sample mean exceeds 21? **(b)** What value will the mean exceed with a probability of 0.6?	*Section 5.11*
7 Give a reason why, although the normal distribution may provide a good model for the weights of new-born mice, it cannot provide an exact model.	*Section 5.9*

5

Test yourself ANSWERS

1 These are unnecessary as the distribution is symmetrical about zero.

2 (a) 1.555; **(b)** −1.405.

3 (a) 0.691; **(b)** 0.0401; **(c)** 0.150; **(d)** 0.615.

4 0.

5 The mean of a large sample will be normally distributed.

6 (a) 0.309; **(b)** 19.5.

7 It is impossible for the mice to have negative weights. A normal distribution would give an infinitesimal but non-zero probability of a baby mouse having a negative weight.

CHAPTER 6

Confidence intervals

Learning objectives

After studying this chapter, you should be able to:

- calculate a confidence interval for the mean of a normal distribution with a known standard deviation
- calculate a confidence interval for the mean of any distribution from a large sample.

6.1 Introduction

Applying statistics often involves using a sample to draw conclusions about a population. This is known as statistical inference. There are two main methods of statistical inference: confidence intervals, which are the subject of this chapter, and hypothesis testing, which is introduced in Unit 2. Confidence intervals are used when we wish to estimate a population parameter and hypothesis testing is used when we wish to make a decision. The calculations involved in the two methods are often similar but the purpose is different.

6.2 Confidence interval for the mean of a normal distribution, standard deviation known

The method is best understood by considering a specific example.

The contents of a large batch of packets of baking powder are known to be normally distributed with standard deviation 7 g. The mean is unknown. A randomly selected packet is found to contain 193 g of baking powder. If this is the only information available the best estimate of the mean contents of the batch is 193 g. This is called a point estimate. However we know that if a different packet had been selected it would almost certainly have contained a different amount of baking powder. It is better to estimate the mean by an interval rather than by a single value. The interval expresses the fact that there is only a limited amount of information and so there is uncertainty in the estimate.

If an observation is taken from a standard normal distribution there is a probability of 0.95 that it will lie in the range ±1.96.

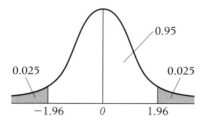

This means that there is a probability of 0.95 that an observation, x, from a normal distribution with mean μ and standard deviation σ will lie in the interval $\mu \pm 1.96\sigma$.

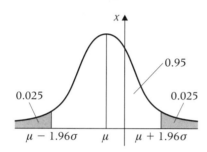

In the example of the packets of baking powder the value of x is known but the value of μ is unknown. If the interval is centred on x, i.e. $x \pm 1.96\sigma$, there is a probability of 0.95 that this interval will contain μ.

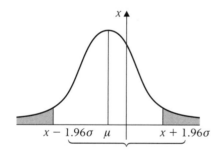

For the packets of baking powder the interval is

$$193 \pm 1.96 \times 7$$

i.e. 193 ± 13.7 or 179.3 to 206.7.

The interval is called a 95% confidence interval. This is because if intervals are calculated in this way then, in the long run, 95% of the intervals calculated will contain the population mean. This also means that 5% will not contain the population mean. Unfortunately there is no way of knowing, in a particular case, whether the interval calculated is one of the 95% which does contain μ or one of the 5% which does not contain μ. You can, however, say that it is much more likely to contain μ than not to contain μ.

> The population mean, μ, is constant but unknown. The interval is known but will be different for each observation, x.

6

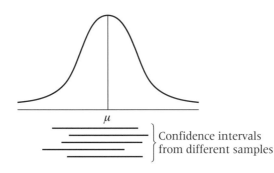

Confidence intervals from different samples

The level of confidence can be increased by widening the interval. For example a 99% confidence interval is

$$193 \pm 2.5758 \times 7$$

i.e. 193 ± 18.0 or 175.0 to 211.0.

It is not possible to calculate a 100% confidence interval.

Intuitively, it seems that a better estimate of μ, the mean contents of the packets of baking powder, will be obtained if we weigh the contents of more than one packet.

Four randomly selected packets were found to contain 193, 197, 212 and 184 g of baking powder. The sample mean is 196.5. The standard deviation of the mean of a sample of size four is $\dfrac{7}{\sqrt{4}} = 3.5$.

A 95% confidence interval for the population mean is

$$196.5 \pm 1.96 \times 3.5$$

i.e. 196.5 ± 6.9 or 189.6 to 203.4.

> This is because there are no limits which contain 100% of a normal distribution.

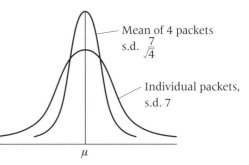

Mean of 4 packets s.d. $\dfrac{7}{\sqrt{4}}$

Individual packets, s.d. 7

μ

Note 1. This interval is half the width of the 95% confidence interval calculated from the weight of a single packet. This has been achieved by increasing the sample size from one to four. However there is very little advantage in increasing a sample of size 21 to one of size 24. To halve the width of the interval you need to **multiply** the sample size by four.

Note 2. If the distribution is not normal the confidence interval will be inaccurate. This could be a major problem for the single observation but would be less serious for the sample of size four. For large samples the sample mean will be approximately normally distributed. Four is not a large sample but the mean of a sample of size four will come closer to following a normal distribution than will the distribution of a single observation.

If $\bar{x}$ is the mean of a random sample of size n from a normal distribution with (unknown) mean μ and (known) standard deviation σ, a $100(1 - \alpha)\%$ confidence interval for μ is given by $\bar{x} \pm z_{\frac{\alpha}{2}} \dfrac{\sigma}{\sqrt{n}}$.

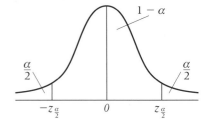

Worked example 6.1 ————————————

A machine fills bottles with vinegar. The volumes of vinegar contained in these bottles are normally distributed with standard deviation 6 ml.

A random sample of five bottles from a large batch filled by the machine contained the following volumes, in millilitres, of vinegar:

986   996   984   990   1002

Calculate a 90% confidence interval for the mean volume of vinegar in bottles of this batch.

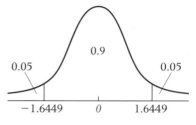

Solution

Sample mean = 991.6 ml
90% confidence interval for mean

$$991.6 \pm 1.6449 \times \frac{6}{\sqrt{5}}$$

i.e. 991.6 ± 4.41 or 987.2 to 996.0 ml

> Answers are usually given to 3 s.f. In this case, because the limits of the interval are close together 4 s.f. is reasonable.

6

Worked example 6.2 ————————————

A food processor produces large batches of jars of jam. In each batch the weight of jam in a jar is known to be normally distributed with standard deviation 7 g. The weights, in grams, of the jam in a random sample of jars from a particular batch were:

481   455   468   457   469   463   469   458

(a) Calculate a 95% confidence interval for the mean weight of jam in this batch of jars.

(b) Assuming the mean weight is at the upper limit of the confidence interval calculated in **(a)**, calculate the limits within which 99% of weights of jam in these jars lies.

(c) The jars are claimed to contain 454 g of jam. Comment on this claim as it relates to this batch.

Solution

(a) Sample mean = 465.0
    95% confidence interval for mean weight of jam is

$$465 \pm 1.96 \times \frac{7}{\sqrt{8}}$$

    i.e. 465 ± 4.851 or 460.1 to 469.9

> Again 4 s.f. is reasonable.

(b) The upper limit of the confidence interval is 469.9.
    If the mean were 469.9, 99% of the weights of jam in the jars would lie in the range

$$469.9 \pm 2.5758 \times 7$$

    i.e. 469.9 ± 18.0 or 451.9 to 487.9.

> Here we are dealing with individual jars, not with sample means.

(c) 454 is below the lower limit of the confidence interval calculated in **(a)**. Hence it is safe to assume that the **mean** weight of jam in the jars exceeds 454 g.

However even if it is assumed that the mean is at the upper limit of the confidence interval, the interval calculated in **(b)** shows that some individual jars will contain less than 454 g.

EXERCISE 6A

1 The potency of a particular brand of aspirin tablets is known to be normally distributed with standard deviation 0.83. A random sample of tablets of this brand was tested and found to have potencies of

      58.7 58.4 59.3 60.4 59.8 59.4 57.7 60.3 61.0 58.2

Calculate:

(a) a 99% confidence interval for the mean potency of these tablets,

(b) a 95% confidence interval for the mean potency of these tablets,

(c) a 60% confidence interval for the mean potency of these tablets.

2 The diastolic blood pressures, in millimetres of mercury, of a population of healthy adults has standard deviation 12.8. The diastolic blood pressures of a random sample of members of an athletics club were measured with the following results:

      79.2   64.6   86.8   73.7   74.9   62.3

(a) Assuming the sample comes from a normal distribution with standard deviation 12.8, calculate:

    (i) a 90% confidence interval for the mean,

    (ii) a 95% confidence interval for the mean,

    (iii) a 99% confidence interval for the mean.

The diastolic blood pressures of a random sample of members of a chess club were also measured with the following results:

      84.6   93.2   104.6   106.7   76.3   78.2

(b) Assuming the sample comes from a normal distribution with standard deviation 12.8, calculate:

    (i) an 80% confidence interval for the mean,

    (ii) a 95% confidence interval for the mean,

    (iii) a 99% confidence interval for the mean.

(c) Comment on the diastolic blood pressure of members of each of the two clubs given that a population of healthy adults would have a mean of 84.8.

3 Applicants for an assembly job are to be given a test of manual dexterity. The times, in seconds, taken by a random sample of applicants to complete the test are

63 229 165 77 49 74 67 59 66 102 81 72 59

Calculate a 90% confidence interval for the mean time taken by applicants. Assume the data comes from a normal distribution with standard deviation 57 s.

4 A rail traveller records the time she has to queue to buy a ticket. A random sample of times, in seconds, were

136   120   67   255   84   99   280   55   78

(a) Assuming the data may be regarded as a random sample from a normal distribution with standard deviation 44 s, calculate a 95% confidence interval for the mean queuing time.

(b) Assume that the mean is at the lower limit of the confidence interval calculated in (a). Calculate limits within which 90% of her waiting times will lie.

(c) Comment on the station manager's claim that most passengers have to queue for less than 25 s to buy a ticket.

5 A food processor produces large batches of jars of pickles. In each batch, the gross weight of a jar is known to be normally distributed with standard deviation 7.5 g. (The gross weight is the weight of the jar plus the weight of the pickles.)
The gross weights, in grams, of a random sample from a particular batch were:

514 485 501 486 502 496 509 491 497
501 506 486 498 490 484 494 501 506
490 487 507 496 505 498 499

(a) Calculate a 90% confidence interval for the mean gross weight of this batch.

The weight of an empty jar is known to be exactly 40 g.

(b) (i) What is the standard deviation of the weight of the pickles in a batch of jars?

   (ii) Assuming that the mean gross weight is at the upper limit of the confidence interval calculated in (a), calculate limits within which 99% of the weights of the pickles would lie.

(c) The jars are claimed to contain 454 g of pickles. Comment on this claim as it relates to this batch of jars.

6.3 Confidence interval for the mean based on a large sample

There are not many real life situations where we wish to use a confidence interval to estimate an unknown population mean when the population standard deviation is known. In most cases where the mean is unknown the standard deviation will also be unknown. If a large sample is available then this will provide a sufficiently good estimate of the standard deviation to enable a confidence interval for the mean to be calculated. The large sample also has the advantage of the sample mean being approximately normally distributed no matter what the distribution of the individual items.

> This could occur in a mass production process where the mean length of components depends on the machine settings but the standard deviation is always the same. However it is unusual.

> If a large random sample is available:
> - it can be used to provide a good estimate of the population standard deviation σ,
> - it is safe to assume that the mean is normally distributed.

> The definition of 'large' is arbitrary. A rule of thumb is that 'large' means at least 30.

Worked example 6.3

Seventy packs of butter, selected at random from a large batch delivered to a supermarket, are weighed. The mean weight is found to be 227 g and the standard deviation is found to be 7.5 g. Calculate a 95% confidence interval for the mean weight of all packs in the batch.

> The sample is large and so it makes little difference whether the divisor n or $n - 1$ is used in calculating the standard deviation. Both will give very similar results. As the population standard deviation is being estimated from a sample it is correct to use the divisor $n - 1$.

Solution

Seventy is a large sample and so although the standard deviation of the weights is not known we may use the standard deviation calculated from the sample. It does not matter whether the distribution is normal or not since the mean of a sample of 70 from any distribution may be modelled by a normal distribution.

The 95% confidence interval for the mean weight of packs of butter in the batch is

$$227 \pm 1.96 \times \frac{7.5}{\sqrt{70}}$$

i.e. 227 ± 1.76 or 225.2 to 228.8.

> At least 4 s.f. are required here. If the answer was rounded to 2 s.f. the interval would disappear completely.

Worked example 6.4

Shoe shop staff routinely measure the length of their customers' feet. Measurements of the length of one foot (without shoes) from each of 180 adult male customers yielded a mean length of 29.2 cm and a standard deviation of 1.47 cm.

(a) Calculate a 95% confidence interval for the mean length of male feet.

(b) Why was it not necessary to assume that the lengths of feet are normally distributed in order to calculate the confidence interval in **(a)**?

(c) What assumption was it necessary to make in order to calculate the confidence interval in **(a)**?

(d) Given that the lengths of male feet may be modelled by a normal distribution, and making any other necessary assumptions, calculate an interval within which 90% of the lengths of male feet will lie.

(e) In the light of your calculations in **(a)** and **(d)**, discuss briefly, the question 'Is a foot a foot long?' (One foot is 30.5 cm.)   [A]

Solution

(a) The 95% confidence interval for the mean length of male feet is

$$29.2 \pm 1.96 \times \frac{1.47}{\sqrt{180}}$$

i.e. 29.2 ± 0.215 or 28.99 to 29.41.

(b) It is not necessary to assume lengths are normally distributed because the central limit theorem states that the mean of a large sample from any distribution will be approximately normally distributed.

(c) To calculate the confidence interval in **(a)** we needed to assume that the sample could be treated as a random sample from the population of all male feet.

(d) 90% of male feet will lie in the interval

$$29.2 \pm 1.6449 \times 1.47$$

i.e. 29.2 ± 2.42 or 26.78 to 31.62.

(e) The confidence interval calculated in **(a)** does not contain 30.5 and so it is very unlikely that the **mean** length of male feet is one foot. The interval calculated in **(d)** does contain 30.5 which indicates that some male feet are a foot long.

EXERCISE 6B

1 A telephone company selected a random sample of size 150 from those customers who had not paid their bills one month after they had been sent out. The mean amount owed by the customers in the sample was £97.50 and the standard deviation was £29.00.
Calculate a 90% confidence interval for the mean amount owed by all customers who had not paid their bills one month after they had been sent out.

2 A sample of 64 fish caught in the river Mirwell had a mean weight of 848 g with a standard deviation of 146 g. Assuming these may be regarded as a random sample of all the fish caught in the Mirwell, calculate, for the mean of this population:

   (a) a 95% confidence interval,

   (b) a 64% confidence interval.

3 A boat returns from a fishing trip holding 145 cod. The mean length of these cod is 74 cm and their standard deviation is 9 cm. The cod in the boat may be regarded as a random sample from a large shoal. The normal distribution may be regarded as an adequate model for the lengths of the cod in the shoal.

   (a) Calculate a 95% confidence interval for the mean length of cod in the shoal.

   (b) It is known that the normal distribution is not a good model for the weights of cod in a shoal. If the cod had been weighed, what difficulties, if any, would arise in calculating a confidence interval for the mean weight of cod in the shoal? Justify your answer.   [A]

4 A sweet shop sells chocolates which appear, at first sight, to be identical. Of a random sample of 80 chocolates, 61 had hard centres and the rest soft centres. The chocolates are in the shape of circular discs and the diameters, in centimetres, of the 19 soft-centred chocolates were:

   2.79 2.63 2.84 2.77 2.81 2.69 2.66 2.71 2.62 2.75
   2.77 2.72 2.81 2.74 2.79 2.77 2.67 2.69 2.75

The mean diameter of the 61 hard-centred chocolates was 2.690 cm.

   (a) If the diameters of both hard-centred and soft-centred chocolates are known to be normally distributed with standard deviation 0.042 cm, calculate a 95% confidence interval for the mean diameter of:

   (i) the soft-centred chocolates,

   (ii) the hard-centred chocolates.

   (b) Calculate an interval within which approximately 95% of the diameters of hard-centred chocolates will lie.

   (c) Discuss, briefly, how useful knowledge of the diameter of a chocolate is in determining whether it is hard- or soft-centred.   [A]

5 Packets of baking powder have a nominal weight of 200 g. The distribution of weights is normal and the standard deviation is 10 g. Average quantity system legislation states that, if the nominal weight is 200 g,

- the average weight must be at least 200 g,
- not more than 2.5% of packages may weigh less than 191 g,
- not more than 1 in 1000 packages may weigh less than 182 g.

A random sample of 30 packages had the following weights:

218 207 214 189 211 206 203 217 183 186
219 213 207 214 203 204 195 197 213 212
188 221 217 184 186 216 198 211 216 200

(a) Calculate a 95% confidence interval for the mean weight.

(b) Assuming that the mean is at the lower limit of the interval calculated in **(a)**, what proportion of packets would weigh,

(i) less than 191 g,

(ii) less than 182 g?

(c) Discuss the suitability of the packets from the point of view of the average quantity system. A simple adjustment will change the mean weight of future packages. Changing the standard deviation is possible but very expensive. Without carrying out any further calculations, discuss any adjustments you might recommend.   [A]

Worked example 6.5

Solid fuel is packed in sacks which are then weighed on scales. It is known that if the full sack weighs μ kg the weight recorded by the scales will be normally distributed with mean μ kg and standard deviation 0.36 kg.

A particular full sack was weighed four times and the weights recorded were 34.7, 34.4, 35.1 and 34.6 kg.

(a) Calculate a 95% confidence interval for the weight of this full sack.

(b) State the width of the interval calculated in **(a)**.

(c) What percentage would be associated with a confidence interval of width 0.3 kg?

(d) How many times would this full sack have to be weighed so that a 95% confidence interval for the weight would be of width 0.3 kg?

Solution

(a) $\bar{x} = 34.7$

95% confidence interval for the mean is

$$34.7 \pm 1.96 \times \frac{0.36}{\sqrt{4}}$$

i.e. 34.7 ± 0.353 or 34.35 to 35.05

(b) width of interval is $2 \times 0.353 = 0.706$

(c) $0.3 = 2z \times \dfrac{0.36}{\sqrt{4}}$

$z = 0.833$

% confidence $= 100\{0.7977 - (1 - 0.7977)\}$
i.e. 59.54 or approximately 60%.

A 60% confidence interval would have width approximately 0.3 kg.

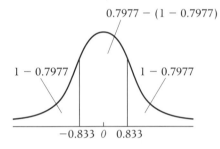

(d) $0.3 = 2 \times 1.96 \times \dfrac{0.36}{\sqrt{n}}$

$\sqrt{n} = 4.704 \qquad n = 22.1.$

If the sack was weighed 22 times a 95% confidence interval would be of width approximately 0.3 kg.

Worked example 6.6

(a) A sample of adult female bears observed in the wild had the following weights in kilograms.

    98 57 71 107 109

The data may be regarded as a random sample from a normally distributed population with a standard deviation of 11 kg.
Calculate a 99% confidence interval for the mean weight of adult female bears.

(b) A sample of adult male bears is also weighed and used to calculate both a 90% and a 95% confidence interval for μ, the mean weight of the population of adult male bears, (i.e. both confidence intervals are calculated from the same sample).

Find the probability that:
(i) the 90% confidence interval does not contain μ,
(ii) the 90% confidence interval does not contain μ but the 95% confidence interval does contain μ,
(iii) the 90% confidence interval contains μ, given that the 95% confidence interval does not contain μ.

(c) Find the probability that the confidence interval calculated in **(a)** does not contain the mean weight of adult female bears and the 90% confidence interval in **(b)** does not contain μ. [A]

Solution

(a) $\bar{x} = 88.4$

99% confidence interval for the mean is

$$88.4 \pm 2.5758 \times \frac{11}{\sqrt{5}}$$

i.e. 88.4 ± 12.7 or 75.7 to 101.1

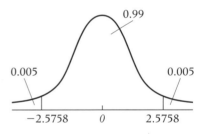

(i) There is a probability of 0.9 that a 90% confidence interval contains μ and so there is a probability of $1 - 0.9 = 0.1$ that a 90% confidence interval does not contain μ.

(ii) If the 95% confidence interval contains μ but the 90% confidence interval does not contain μ then $\bar{x}$ must lie in the shaded area.

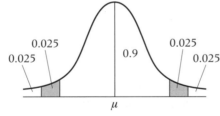

Probability is 0.05.

(iii) If the 95% confidence interval does not contain μ it is impossible for the 90% confidence interval to contain μ.

The probability is 0.

(c) The probability that the 99% confidence interval calculated in **(a)** does not contain the mean weight of female bears is $1 - 0.99 = 0.01$.

The probability the 90% confidence interval calculated in **(b)** does not contain μ is $1 - 0.9 = 0.1$.

Assuming that these two samples are independent the required probability is $0.01 \times 0.1 = 0.001$.

EXERCISE 6C

1 A random sample of experimental components for use in aircraft engines was tested to destruction under extreme conditions. The survival times, X days, of ten components were as follows:

   207   381   111   673   234   294   897   144   418   554

   (a) Assuming that the survival time, under these conditions, for all the experimental components is normally distributed with standard deviation 240 days, calculate a 90% confidence interval for the mean of X.

   (b) State the probability that the confidence interval calculated in **(a)** does not contain the mean of X.

2 A car manufacturer purchases large quantities of a particular component. The working lives of the components are known to be normally distributed with mean 2400 hours and standard deviation 650 hours. The manufacturer is concerned about the large variability and the supplier undertakes to improve the design so that the standard deviation is reduced to 300 hours.

   A random sample of five of the new components is tested and found to last

   2730   3120   2980   2680   2800 hours.

   Assuming that the lives of the new components are normally distributed with standard deviation 300 hours:

   (a) **(i)** Calculate a 90% confidence interval for their mean working life.
   (ii) How many of the new components would it be necessary to test in order to make the width of a 95% confidence interval for the mean just less than 100 hours?

   (b) Lives of components commonly follow a distribution which is not normal. If the assumption of normality is invalid, comment briefly on the amount of uncertainty in your answers to **(a)(i)** and **(ii)**.

   (c) Is there any reason to doubt the assumption that the standard deviation of the lives of the new components is 300 hours? [A]

3 A supermarket sells cartons of tea bags. The weight, in grams, of the contents of the cartons in any batch is known to be normally distributed with mean μ_T and standard deviation 4. In order to compare the actual contents with that claimed by the supplier, a manager weighed the contents of a random sample of five cartons from a large batch and obtained the following results, in grams:

196   202   198   197   190

(a) Calculate:
    (i) a 95% confidence interval for μ_T,
    (ii) a 60% confidence interval for μ_T.

The manager intends to do the same thing tomorrow (i.e. to weigh the contents of a random sample of cartons of tea and to use the data collected to calculate both a 95% and a 60% confidence interval).

(b) State the probability that:
    (i) the 95% confidence interval she calculates will not contain μ_T,
    (ii) neither of the confidence intervals she calculates will contain μ_T.

The manager also intends to weigh the contents of jars of coffee from a batch in order to calculate a 95% and a 60% confidence interval for the mean contents, μ_C, of jars in the batch. However, in this case, the 95% confidence interval will be calculated from one random sample and the 60% confidence interval calculated from a second, independent, random sample.

(c) Find the probability that neither the 60% nor the 95% confidence interval for the mean contents of jars of coffee will contain μ_C. [A]

4 Batteries supplied to a large institution for use in electric clocks had a mean working life of 960 days with a standard deviation of 135 days.

A sample from a new supplier had working lives of

1020, 998, 894, 921, 843, 1280, 1302, 782, 694, 1350 days.

Assume that the data may be regarded as a random sample from a normal distribution with standard deviation 135 days.

(a) For the working lives of batteries from the new supplier, calculate a 95% confidence interval for the mean.

(b) The institution would like batteries with a large mean. Compare the two sources of supply.

(c) State the width of the confidence interval calculated in **(a)**.

(d) What percentage would be associated with an interval of width 100 days calculated from the data above?

(e) How large a sample would be needed to calculate a 90% confidence interval of width approximately 100 days?

[A]

5 A manufacturer makes batteries for use in bicycle lights. The working lives, in hours, of the batteries are known to be normally distributed with a standard deviation of 1.8. A random sample of batteries was tested and their working lives were as follows:

48.2   49.6   47.1   50.0   46.8   47.2   47.9

(a) Calculate a 95% confidence interval for the mean working life of the batteries.

(b) State the width of the confidence interval you have calculated.

(c) What percentage would be associated with a confidence interval of width 2 hours calculated from the given data?

(d) A further random sample of size seven is to be taken and used to calculate a confidence interval of width 2 hours. State the probability that this confidence interval will not contain the mean working life. [A]

6 A firm is considering providing an unlimited supply of free bottled water for employees to drink during working hours. To estimate how much bottled water is likely to be consumed, a pilot study is undertaken. On a particular day-shift, ten employees are provided with unlimited bottled water. The amount each one consumes is monitored. The amounts, in millilitres, consumed by these ten employees are as follows:

110   0   640   790   1120   0   0   2010   830   770

(a) Assuming the data may be regarded as a random sample from a normal distribution with standard deviation 510, calculate a 95% confidence interval for the mean amount consumed on a day-shift.

(b) (i) Give a reason, based on the data collected, why the normal distribution may not provide a suitable model for the amount of free bottled water which would be consumed by employees of the firm.

(ii) A normal distribution may provide an adequate model but cannot provide an exact model for the amount of bottled water consumed. Explain this statement, giving a reason which does not depend on the data collected.

(c) Following the pilot study the firm offers the bottled water to all the 135 employees who work on the night-shift. The amounts they consume on the first night have a mean of 960 ml with a standard deviation of 240 ml.

(i) Assuming these data may be regarded as a random sample, calculate a 90% confidence interval for the mean amount consumed on a night-shift.

(ii) Explain why it was not necessary to know that the data came from a normal distribution in order to calculate the confidence interval in **(c)(i)**.

(iii) Give one reason why it may be unrealistic to regard the data as a random sample of the amounts that would be consumed by all employees if the scheme was introduced on all shifts on a permanent basis.

[A]

7 A health food cooperative imports a large quantity of dates and packs them into plastic bags labelled 500 g. Georgina, a Consumer Protection Officer, checked a random sample of 95 bags and found the contents had a mean weight of 498.6 g, and a standard deviation of 9.3 g.

(a) Assuming that weights follow a normal distribution, calculate, for the mean weight of contents of all the bags:
   (i) a 95% confidence interval,
   (ii) an 80% confidence interval.

(b) The health food cooperative also imports raisins. Georgina intends to take a random sample of 500 g packets of raisins, weigh the contents and use the results to calculate an 80% and a 95% confidence interval for μ, the mean weight of the contents of all the cooperative's packets of raisins.
   (i) Find the probability that:
      (A) the 80% confidence interval contains μ,
      (B) the 95% confidence interval contains μ but the 80% confidence interval does not.
   (ii) Instead of calculating both confidence intervals from the same sample, Georgina now decides to calculate the 95% confidence interval from one sample and the 80% confidence interval from a second independent random sample. Find the probability that the 95% confidence interval contains μ, but the 80% confidence interval does not.

[A]

8 Applicants to join a police force are tested for physical fitness. Based on their performance, a physical fitness score is calculated for each applicant. Assume that the distribution of scores is normal.

(a) The scores for a random sample of ten applicants were

   55   23   44   69   22   45   54   72   34   66

Experience suggests that the standard deviation of scores is 14.8.

Calculate a 99% confidence interval for the mean score of all applicants.

(b) The scores of a further random sample of 110 applicants had a mean of 49.5 and a standard deviation of 16.5. Use the data from this second sample to calculate:
  (i) a 95% confidence interval for the mean score of all applicants,
  (ii) an interval within which the score of approximately 95% of applicants will lie.

(c) By interpreting your results in **(b)(i)** and **(b)(ii)**, comment on the ability of the applicants to achieve a score of 25.

(d) Give two reasons why a confidence interval based on a sample of size 110 would be preferable to one based on a sample of size 10.

(e) It is suggested that a much better estimate of the mean physical fitness of all applicants could be made by combining the two samples before calculating a confidence interval. Comment on this suggestion.   [A]

Key point summary

1 If $\bar{x}$ is the mean of a random sample of size n from a   *p108*
normal distribution with (unknown) mean μ and (known) standard deviation σ, a $100(1 - \alpha)\%$
confidence interval for μ is given by $\bar{x} \pm z_{\frac{\alpha}{2}} \dfrac{\sigma}{\sqrt{n}}$.

2 If a large random sample is available: *p112*
  • it can be used to provide a good estimate of the population standard deviation σ,
  • it is safe to assume that the mean is normally distributed.

Test yourself	**What to review**
1 The lengths of components produced by a machine are normally distributed with standard deviation 0.005 cm. A random sample of components measured 1.002 1.007 1.016 1.009 1.003 cm. Calculate a 95% confidence interval for the population mean.	*Section 6.2*
2 State the probability that an 85% confidence interval for μ does not contain μ.	*Section 6.2*
3 Comment on the claim that the mean length of the components in question 1 is 1.011 cm.	*Section 6.2*
4 The contents of a random sample of 80 tins of vindaloo cooking sauce from a large batch were weighed. The sample mean content was 284.2 g and the standard deviation, s, was found to be 4.1 g. Calculate a 95% confidence interval for the population mean.	*Section 6.3*
5 How would your answer to question 4 be affected if it was later discovered that the batch contained tins which had been produced on two different machines and the distribution of the weights was bimodal?	*Section 6.3*
6 How would your answer to question 4 be affected if it was later discovered that the sample was not random?	*Section 6.3*

6

Test yourself **ANSWERS**

1 1.0030–1.0118 cm.

2 0.15.

3 Although 1.011 is higher than the sample mean it lies within the confidence interval. There is therefore no convincing evidence that the mean is not 1.011 as claimed.

4 283.30–285.10 g.

5 It would not affect the answer. The sample is stated to be a random sample from the whole batch and the central limit theorem applies to all distributions whether bimodal or not.

6 If the sample was not random this would make the confidence interval unreliable. For example, the sample of tins could all have come from the same machine leading to a biased result.

CHAPTER 7

Correlation

Learning objectives

In earlier chapters, only single variables have been considered. Now you will be working with pairs of variables.

After studying this chapter, you should be able to:

■ investigate the strength of a linear relationship between two variables by using suitable statistical analysis
■ evaluate and interpret the product moment correlation coefficient.

7.1 Scatter diagrams

Scatter diagrams are used where we are examining possible relationships between two variables.

An athlete, recovering from injury, had her pulse rate measured after performing a predetermined number of step-ups in a gymnasium. The measurements were made at weekly intervals. The table below shows the number of step-ups, x, the pulse rate, y beats per minute, and the week in which the measurement was made.

Week	1	2	3	4	5	6	7	8
x	15	50	35	25	20	30	10	45
y	114	155	132	112	96	105	78	113

To construct a scatter diagram you simply plot points with coordinates (x, y) for each of the 8 weeks.

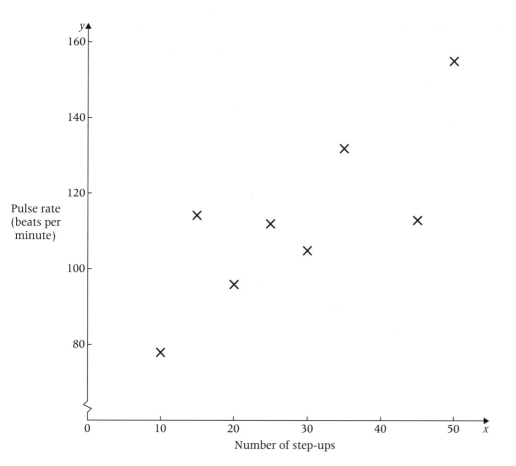

In this case, as you would expect, there appears to be a clear tendency for the pulse rate to increase with the number of step-ups.

7.2 Interpreting scatter diagrams

Interpreting a scatter diagram is often the easiest way for you to decide whether correlation exists. Correlation means that there is a linear relationship between the two variables. This could mean that the points lie on a straight line, but it is much more likely to mean that they are scattered about a straight line.

7

The main types of scatter diagram

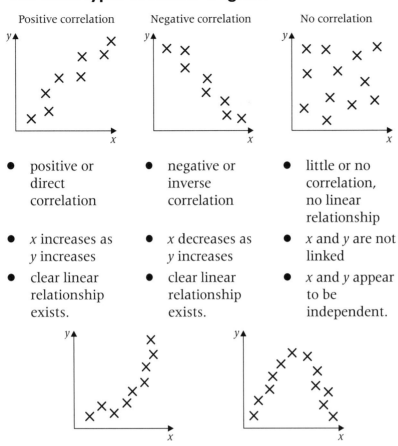

Positive correlation

- positive or direct correlation

- *x* increases as *y* increases
- clear linear relationship exists.

Negative correlation

- negative or inverse correlation

- *x* decreases as *y* increases
- clear linear relationship exists.

No correlation

- little or no correlation, no linear relationship

- *x* and *y* are not linked
- *x* and *y* appear to be independent.

- *x* and *y* are clearly linked by a non-linear relationship.

7.3 Studying results

The table below gives the marks obtained by ten pupils taking maths and physics tests.

Pupil	A	B	C	D	E	F	G	H	I	J
Maths mark (out of 30) *x*	20	23	8	29	14	11	11	20	17	17
Physics mark (out of 40) *y*	30	35	21	33	33	26	22	31	33	36

Is there a connection between the marks obtained by the ten pupils in the maths and physics tests?

The starting point would be to plot the marks on a scatter diagram.

The areas in the bottom-right and top-left of the graph are almost empty so there is a clear tendency for the points to run from bottom-left to top-right. This indicates that positive correlation exists between *x* and *y*.

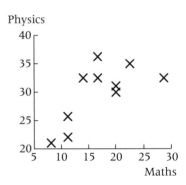

Calculating the means:

$$\bar{x} = \frac{170}{10} = 17$$

and

$$\bar{y} = \frac{300}{10} = 30.$$

Using these lines, the graph can be divided into four regions to show this tendency very clearly.

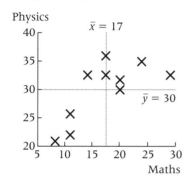

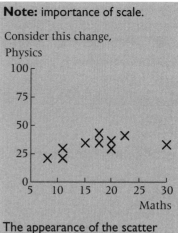

Note: importance of scale.

Consider this change,

The appearance of the scatter diagram is now very different. The existence of correlation is much more difficult to identify. Scales should cover the range of the given data.

The table below gives the marks obtained by the ten pupils taking maths and history tests.

Pupil	A	B	C	D	E	F	G	H	I	J
Maths mark (out of 30) x	20	23	8	29	14	11	11	20	17	17
History mark (out of 60) z	28	21	42	32	44	56	36	24	51	26

Calculating the mean for z:

$$\bar{z} = \frac{360}{10} = 36$$

The scatter diagram for maths and history shows a clear tendency for points to run from top-left to bottom-right. This indicates that negative correlation exists between x and z.

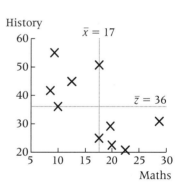

7.4 Product moment correlation coefficient (PMCC)

(This is often known as **Pearson's correlation coefficient** after **Karl Pearson**, an applied mathematician who worked on the application of statistics to genetics and evolution.)

How can the strength of correlation be quantified?

There are two main points to consider.

- How close to a straight line are the points?
- Is the correlation positive or negative?

The product moment correlation coefficient, r, gives a standardised measure of correlation which can be used for comparisons between different sets of data.

S_{xx}, S_{yy} and S_{xy} are used to evaluate r where:

$S_{xx} = \sum(x_i - \bar{x})^2$, $S_{yy} = \sum(y_i - \bar{y})^2$ and

$S_{xy} = \sum(x_i - \bar{x})(y_i - \bar{y})$

r is given by $\dfrac{S_{xy}}{\sqrt{S_{xx}S_{yy}}}$

These formulae are given in the AQA formulae book.

Formula

The computational form of this equation which is most commonly used is:

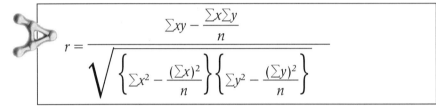

$$r = \frac{\sum xy - \dfrac{\sum x \sum y}{n}}{\sqrt{\left\{\sum x^2 - \dfrac{(\sum x)^2}{n}\right\}\left\{\sum y^2 - \dfrac{(\sum y)^2}{n}\right\}}}$$

r is obtainable directly from all calculators with regression facility. This is recommended in the exam.

Sketches to illustrate examples of possible values of r.

Values of r

Some worked examples

Returning to the maths and physics marks in section 8.2.

To illustrate the calculation involved in evaluating r, the following additional summations are needed:

$\sum x^2 = 3250$, $\sum y^2 = 9250$, $\sum xy = 5313$.

You can then see that

$$S_{xx} = 3250 - \frac{170^2}{10} = 360$$

and

$$S_{yy} = 9250 - \frac{300^2}{10} = 250$$

Then, $S_{xy} = 5313 - \dfrac{170 \times 300}{10} = 213$

So

$$r = \frac{213}{\sqrt{360 \times 250}} = 0.71$$

This, of course, can be found directly from your calculator.

The interpretation of the value of r is very important. The value of r tells you how close the points are to lying on a straight line.

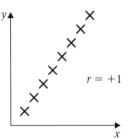

$r = +1$

Exact positive correlation

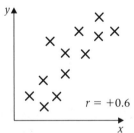

$r = +0.6$

Weak positive correlation

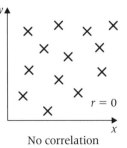

$r = 0$

No correlation

It is always true that:

$$-1 \leq r \leq +1$$

$r = +1$ indicates **ALL the points lie on a line** with positive gradient

$r = -1$ indicates **ALL the points lie on a line** with negative gradient

$r = 0$ indicates that there is **no linear connection** at all between the two sets of data.

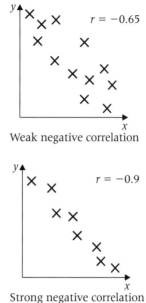

Weak negative correlation

$r = -0.65$

$r = -0.9$

Strong negative correlation

The value obtained in this example, $r = 0.71$, would indicate a fairly strong positive correlation between the test score in maths and the test score in physics.

Worked example 7.1

A group of 12 children participated in a psychological study designed to assess the relationship, if any between age, x years, and average total sleep time (ATST), y minutes. To obtain a measure for ATST, recordings were taken on each child on five consecutive nights and then averaged. The results are below.

Child	Age x (years)	ATST y (minutes)
A	4.4	586
B	6.7	565
C	10.5	515
D	9.6	532
E	12.4	478
F	5.5	560
G	11.1	493
H	8.6	533
I	14.0	575
J	10.1	490
K	7.2	530
L	7.9	515

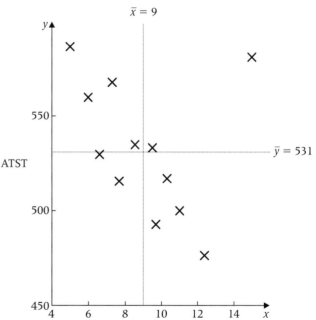

Calculate the product moment correlation coefficient between x and y and interpret your result.

Solution

$\sum x = 108$ and $\sum y = 6372$

$\sum x^2 = 1060.1$, $\sum y^2 = 3\,396\,942$

$\sum xy = 56\,825.4$

$$S_{xx} = 1060.1 - \frac{108^2}{12} = 88.1$$

$$S_{yy} = 3\,396\,942 - \frac{6372^2}{12} = 13\,410$$

Then

$$S_{xy} = 56\,825.4 - \frac{108 \times 6372}{12} = -522.6$$

So

$$r = \frac{-522.6}{\sqrt{88.1 \times 13\,410}} = -0.481 \text{ (to 3 s.f.)}$$

Considering the value of r and the scatter diagram, there is evidence of weak negative correlation between age and ATST. This would indicate that older children have less ATST than younger children. However, the relationship is fairly weak.

> Note that it would be worth investigating child *I* who seems to have an abnormally high ATST. Perhaps the child was ill during the experiment or perhaps there is some other reason for the excessive amount of sleep.

Worked example 7.2

The following data indicate the level of sales for ten models of pen sold by a particular company. The sales, together with the selling price of the pen, are given in the table below.

Model	Price, x (£)	Sales, y (00s)
A	2.5	30
B	5	35
C	15	25
D	20	15
E	7.5	25
F	17.5	10
G	12	15
H	6	20
I	25	8
J	30	10

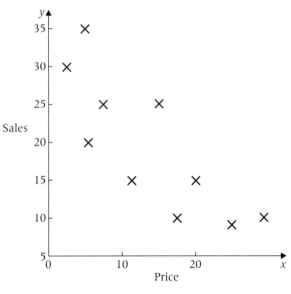

$\sum x = 140.5$ $\sum y = 193$
$\sum x^2 = 2723.75$ $\sum y^2 = 4489$ $\sum xy = 2087.5$

Plot these data on a scatter diagram. Evaluate the product moment correlation coefficient and interpret your answers with reference to the data supplied.

> Note that care must be taken not to approximate prematurely in calculations or else r may be inaccurate.

Solution

$$S_{xx} = 2723.75 - \frac{140.5^2}{10} = 749.725$$

$$S_{yy} = 4489 - \frac{193^2}{10} = 764.1$$

Then,

$$S_{xy} = 2087.5 - \frac{140.5 \times 193}{10} = -624.15$$

So

$$r = \frac{-624.15}{\sqrt{749.725 \times 764.1}} = -0.825 \text{ (to 3 s.f.)}$$

Considering the value of r and the scatter diagram, there is evidence of quite strong negative correlation between x and y.

This would indicate that there are fewer sales of the more expensive pens and this trend follows a linear relationship.

Using prematurely rounded figures:

$S_{xx} = 750$ and $S_{yy} = 764$
$S_{xy} = -624$

$$r = \frac{-624}{\sqrt{750 \times 764}}$$

$$= \frac{-624}{757} = -0.824 \text{ (to 3 s.f.)}$$

An error has now occurred. It is only the **final answer** which should be rounded to three significant figures.

7.5 Limitations of correlation

It is very important to remember a few key points about correlation.

Remember it is recommended that r is obtained directly using a calculator.

Non-linear relationships

As illustrated in section 7.2, r measures linear relationships only. It is of no use at all when a non-linear relationship is evident. There may well be a very clear relationship between the variables being considered but if that relationship is not linear then r will not help at all.

Note that clear non-linear relationships identified on scatter diagrams should always be commented upon but the evaluation of r is not appropriate.

The scatter diagram should reveal this.

Cause and effect

A student does some research in a primary school and discovers a very strong direct correlation between length of left foot and score in a mental maths test. Does this mean that stretching a child's foot will make them perform better in maths?

Clearly this is ridiculous and the probable hidden factor is age: older children have bigger feet and a better ability at maths.

Note that any suggestion that correlation may indicate cause and effect in the relationship between two variables should be considered very carefully!

The correlation found between foot length and score in maths is often called *spurious* and should be treated with caution.

Age ⟨ Foot length / Mental maths score

Freak results

An unusual result can drastically alter the value of r. Unexpected results should always be commented upon and investigated further as their inclusion or exclusion in any calculations can completely change the final result.

Imagine the effect on *r* if the point P is to be removed from correlation calculations using the data below.

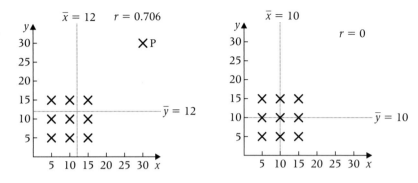

Worked example 7.3 —————————

Plot scatter diagrams on separate axes for the following data sets:

(a)

x	15	10	5	20	25	10	25	10
y	3	2.5	5	5	4	5	5	3

(b)

x	2.5	2.8	3	3.2	4.5	5	6	8
y	20	14	10	8	6	4	3	2

It has been suggested that the product moment correlation coefficient should be evaluated for both sets of data. By careful examination of your scatter diagrams, comment on this suggestion in each case.

Solution

(a) The scatter diagram indicates little or no correlation between the two variables. *r* could be evaluated but would clearly be close to zero.

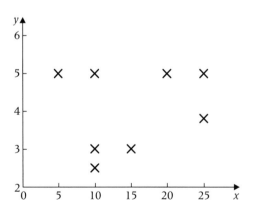

(b) *r* may well indicate fairly strong negative correlation between the two variables **but** the scatter diagram clearly shows that the relationship is non-linear and hence *r* is an inappropriate measure.

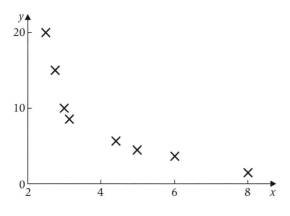

Worked example 7.4

A tasting panel was asked to assess biscuits baked from a new recipe. Each member was asked to assign a score from 0 to 100 for texture (x), for flavour (y) and for sweetness (z).

The scores assigned by the ten tasters were as follows:

Taster	1	2	3	4	5	6	7	8	9	10
x	43	59	76	28	53	55	81	49	38	47
y	67	82	75	48	91	63	67	51	44	54

(a) Draw a scatter diagram to illustrate the data.

(b) Calculate the value of the product moment correlation coefficient between x and y.

(c) State, briefly, how you would expect the scatter diagram to alter if the tasters were given training in how to assign scores before the tasting took place.

(d) Given that $\Sigma(z_i - \bar{z})^2 = 2516.9$ and $\Sigma(y_i - \bar{y})(z_i - \bar{z}) = 1979.8$, calculate the product moment correlation coefficient between y and z.

Solution

(a)

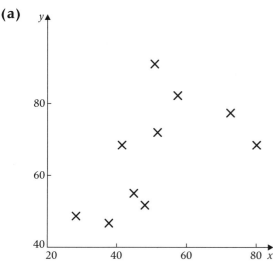

(b) $\sum x = 529, \bar{x} = 52.9, \sum y = 642, \bar{y} = 64.2$

$\sum xy = 35\,187, \sum x^2 = 30\,339, \sum y^2 = 43\,334$

$$S_{xx} = 30\,339 - \frac{529^2}{10} = 2354.9$$

and $S_{yy} = 43\,334 - \dfrac{642^2}{10} = 2117.6$

and $S_{xy} = 35\,187 - \dfrac{529 \times 642}{10} = 1225.2$

therefore, $r = \dfrac{1225.2}{\sqrt{2354.9 \times 2117.6}} = 0.549$ (to 3 s.f.)

> Note that a calculator can be used to obtain r directly.

(c) The scores would be less variable.

The scatter diagram would be more compact but the overall shape would be similar.

> Training would lead to a more consistent scale for x and y. Without training, people's views on texture or flavour would vary widely.

(d) $r = \dfrac{\sum(y_i - \bar{y})(z_i - \bar{z})}{\sqrt{\sum(y_i - \bar{y})^2 \sum(z_i - \bar{z})^2}}$

$= \dfrac{S_{yz}}{\sqrt{S_{yy}S_{zz}}},$

therefore,

$$r = \frac{1979.8}{\sqrt{2117.6 \times 2516.9}} = 0.858$$

> A calculator cannot be used to obtain r directly in this case – the formula must be used.

Worked example 7.5

The following data show the annual income per head, x ($US), and the infant mortality, y (per thousand live births), for a sample of 11 countries.

Country	x	y
A	130	150
B	5950	43
C	560	121
D	2010	53
E	1870	41
F	170	169
G	390	143
H	580	59
I	820	75
J	6620	20
K	3800	39

$\sum x = 22\,900,$ $\sum x^2 = 102\,724\,200,$

$\sum y = 913,$ $\sum y^2 = 103\,517,$ $\sum xy = 987\,130.$

(a) Draw a scatter diagram of the data. Describe the relationship between income per head and infant mortality suggested by the diagram.

(b) An economist asks you to calculate the product moment correlation coefficient.

  (i) Carry out this calculation.

  (ii) Explain briefly to the economist why this calculation may not be appropriate.

Solution

(a)

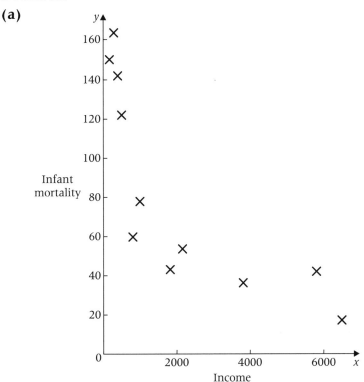

Infant mortality appears to decline as income per head increases.

The decrease is not uniform but is much more marked for the very low incomes than for the higher income countries.

(b) **(i)** $S_{xx} = 102\,724\,200 - \dfrac{22\,900^2}{11} = 55\,050\,563.64$

and $S_{yy} = 103\,517 - \dfrac{913^2}{11} = 27\,738$

and $S_{xy} = 987\,130 - \dfrac{22\,900 \times 913}{11} = -913\,570$

therefore, $r = \dfrac{-913\,570}{\sqrt{55\,050\,563.64 \times 27\,738}} = -0.739$

(to 3 s.f.)

Note that this can be found directly from a calculator.

7

(ii) PMCC measures the strength of a linear relationship. It is not a suitable measure for data which clearly shows a non-linear relationship, as in this case.

> Look back to the beginning of section 7.5. A clear curve is seen.

EXERCISE 7A

1 (a) For each of the following scatter diagrams, state whether or not the product moment correlation coefficient is an appropriate measure to use.

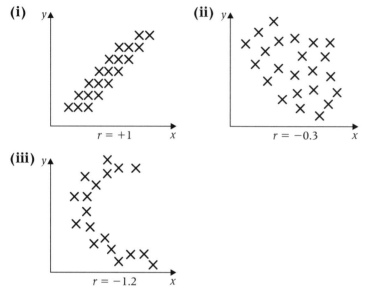

(i) $r = +1$

(ii) $r = -0.3$

(iii) $r = -1.2$

(b) State, giving a reason, whether or not the value underneath each diagram might be a possible value of this correlation coefficient.

2 Estimate, **without undertaking any calculations**, the product moment correlation coefficient between the variables in each of the scatter diagrams given:

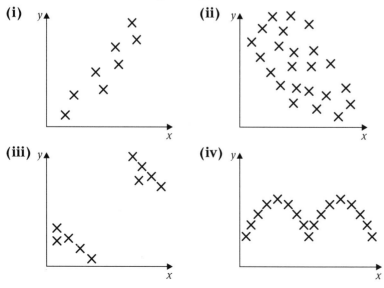

(i)

(ii)

(iii)

(iv)

3 For each of the following sets of data:

(a) draw a scatter diagram,

(b) calculate the product moment correlation coefficient between x and y.

(i)

x	1	3	6	10	12
y	5	13	25	41	49

(ii)

x	1	3	6	10	12
y	44	34	24	14	4

(iii)

x	1	1	3	5	5
y	5	1	3	1	5

(iv)

x	1	3	6	9	11
y	12	28	37	28	12

4 The diameters of the largest lichens growing on gravestones were measured.

Age of gravestone (x years)	Diameter of lichen (y mm)
9	2
18	3
20	4
31	20
44	22
52	41
53	35
61	22
63	28
63	32
64	35
64	41
114	51
141	52

(a) Plot a scatter diagram to show the data.

(b) Calculate the values of $\bar{x}$ and $\bar{y}$ and show these as vertical and horizontal lines.

(c) Find the values of S_{xx}, S_{yy} and r.

5 A metal rod was gradually heated and its length, L, and temperature, T, were measured several times.

Temperature, T (°C)	Length, L (cm)
14	100.0
21	103.8
24	100.6
29	111.0
37	116.1
40	119.9

7

(a) Draw a scatter diagram to show the data, plotting L against T.

(b) Find the value of r, the product moment correlation coefficient.

(c) It is suspected that a major inaccuracy may have occurred in one or more of the recorded values. Discard any readings which you consider may be untrustworthy and find the new value for r.

Comment on your results.

6 In a workshop producing hand-made goods a score is assigned to each finished item on the basis of its quality (the better the quality the higher the score). The number of items produced by each of 15 craftsmen on a particular day and their average quality score are given below.

Craftsman	No. of items produced, x	Average quality score, y
1	14	6.2
2	23	7.3
3	17	4.9
4	32	7.1
5	16	5.2
6	19	5.7
7	17	5.9
8	25	6.4
9	27	7.3
10	31	6.1
11	17	5.4
12	18	5.7
13	26	6.9
14	24	7.2
15	22	4.8

(a) Draw a scatter diagram to show the data.

(b) Calculate the product moment correlation coefficient between x and y.

(c) The owner of the firm believes that the quality of the output is suffering because some of the craftsmen are working too fast in order to increase bonus payments. Explain to him the meaning of your results, and state what evidence, if any, they provide for or against his belief.

7 During the summer of 1982 the National Leisure Council, on behalf of the Government, conducted a survey into all aspects of the nation's leisure time. The table shows the amount spent per month on sporting pastimes and the total amount spent per month on all leisure activities for a random sample of 13 young married men.

Man	Amount on sport, x	Total amount, y
A	9.0	50.1
B	4.2	46.6
C	12.9	52.4
D	6.1	45.1
E	14.0	56.3
F	1.5	46.6
G	17.4	52.0
H	10.2	48.7
I	18.1	56.0
J	2.9	48.0
K	11.6	54.1
L	15.2	53.3
M	7.3	51.7

(a) Draw a scatter diagram for this data.

(b) Calculate the product moment correlation coefficient for the data.

(c) Comment, with reasons, upon the usefulness, or otherwise, of the above correlation analysis.

8 A clothing manufacturer collected the following data on the age, x months, and the maintenance cost, y (£), of his sewing machines.

Machine	Age, x	Cost, y
A	13	24
B	75	144
C	64	110
D	52	63
E	90	240
F	15	20
G	35	40
H	82	180
I	25	42
J	46	50
K	50	92

(a) Plot a scatter diagram of the data.

(b) Calculate the product moment correlation coefficient.

(c) Comment on your result in **(b)** by making reference to the scatter diagram drawn in **(a)**.

9 The following data relate to a random sample of 15 males, all aged between 40 and 60 years. The measurements given are the level of heart function (out of 100), the percentage of baldness and the average number of hours spent watching television each day.

Male	Heart function	Baldness (%)	Hours of TV
1	42	83	6.2
2	65	66	2.2
3	86	32	1.8
4	32	74	8.3
5	56	69	7.6
6	48	74	6.5
7	92	25	0.8
8	78	30	5.9
9	68	32	2.2
10	52	54	4.4
11	53	58	4.6
12	69	76	2.7
13	57	63	5.8
14	89	38	0.2
15	65	41	4.6

(a) Calculate the value of the product moment correlation coefficient between heart function and percentage baldness.

(b) Calculate the value of the product moment correlation coefficient between heart function and average number of hours of television watched per day.

(c) Comment on the values of the correlation coefficients found in **(a)** and **(b)** and interpret your results.

(d) Do you consider that males aged between 40 and 60 should be advised to reduce the number of hours that they spend watching television in order to ensure a better heart function? Explain your answer.

Key point summary

1 A scatter diagram should be drawn to judge
whether correlation is present.   *p124*

2 The product moment correlation coefficient,   *p128, 129, 131*

$$r = \frac{\sum xy - \frac{\sum x \sum y}{n}}{\sqrt{\left\{\sum x^2 - \frac{(\sum x)^2}{n}\right\}\left\{\sum y^2 - \frac{(\sum y)^2}{n}\right\}}} \quad \text{or} \quad \frac{S_{xy}}{\sqrt{S_{xx}S_{yy}}}$$

Remember, this can be found directly using a
calculator.

r is a measure of **linear** relationship only and
$-1 \le r \le +1$

Do not refer to *r* if a scatter diagram clearly shows a
non-linear connection.

3 $r = +1$ or $r = -1$ implies that the points all **exactly**   *p129*
lie on a **straight line**.

$r = 0$ implies **no** linear relationship is present.

But ... no linear relationship between the variables
does not necessarily mean that $r = 0$.

4 Even if *r* is close to $+1$ or -1, **no causal link** should   *p131*
be assumed between the variables without thinking
very carefully about the nature of the data involved.

Remember the feet stretching! Will it really help you
to get better at maths?

7

Test yourself	What to review

1 Which of the following could not be a value for a product moment correlation coefficient?

Section 7.4

   (a) $r = 0.98$,

   (b) $r = -0.666$,

   (c) $r = 1.2$,

   (d) $r = 0.003$.

2 Which of the following scatter diagrams has a corresponding product moment correlation coefficient given which is not appropriate?

Section 7.4

   (a) $r = -0.86$

   (b) $r = 0.784$

   (c) $r = -0.145$

Test yourself (*continued*)	What to review

3 For the following data, plot a scatter diagram and evaluate the product moment correlation coefficient.   *Section 7.4*

x	8	6	5	2	-1	-3	-6
y	-9	-8	-8	0	-5	2	7

4 Explain the meaning of spurious correlation with reference to the following statement: *Section 7.4*

'Between 1988 and 1998, the product moment correlation coefficient between the number of incidents of violent juvenile offences taken to court each year and the average number of hours per week which 16- to 19-year-olds spent watching television was found to be 0.874, indicating a high level of correlation.'

5 The weight losses for ten females enrolled on the same Watch and Weight course at a local Sports Centre are given below.   *Section 7.4*

Weeks on course	Weight loss (kg)
5	7.6
15	23
12	19.6
3	1.2
10	17.4
8	15.2
20	25.5
10	14
5	2.4
8	9.5

(a) Plot a scatter diagram.

(b) Evaluate the product moment correlation coefficient and comment on its value, referring also to the scatter diagram.

7

1 (c).

2 (c).

3 $r = -0.904$.

Graph of y against x

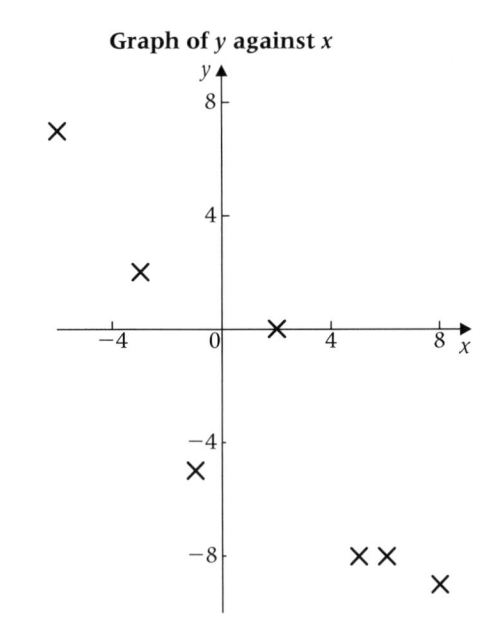

4 Spurious refers to the fact that the link between the two variables may not be causal. They may be two effects from a different cause.

5 (a) **Graph of weight loss against time on course**

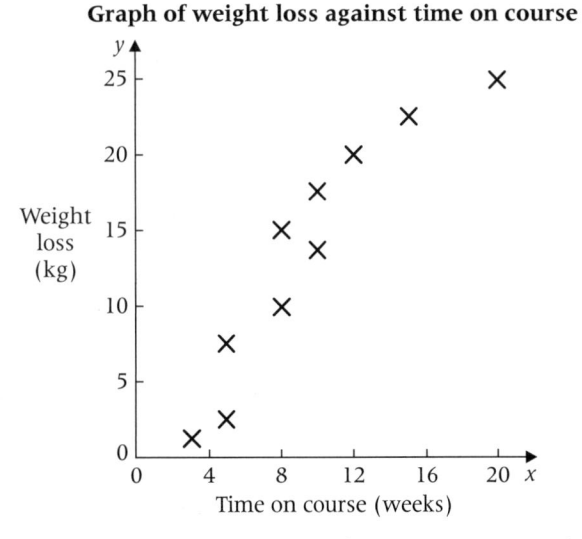

(b) $r = 0.936$. Strong positive correlation. Some suggestion from scatter diagram that weight loss is reaching a peak.

CHAPTER 8

Regression

Learning objectives

This chapter continues with the analysis of bivariate data that was started in chapter 7.

After studying this chapter, you should be able to:
- find the equation of regression lines using the method of least squares
- interpret the values obtained for the gradient and intercept of your regression line
- plot a regression line on a scatter diagram and use the line for prediction purposes
- calculate residuals and, when appropriate, use them to check the fit of a regression line and to improve predictions.

8.1 What is regression analysis?

In linear regression analysis, bivariate data are first examined by drawing a scatter diagram in order to determine whether a linear relationship exists (by eye or by finding the product moment correlation coefficient or PMCC). Then the actual equation of the line of best fit is obtained in the form:

> $$y = a + bx$$
> This is called the
> **line of y on x.**

This equation may then be used to predict a value of y from a given value of x.

> The regression line is often called the **line of best fit**.

> Remember that we are still considering linear relationships (**straight lines**) only.

> Look back to section 7.2.

> On the scatter diagram, y is plotted on the vertical axis and x on the horizontal.

> In pure maths, you may be more familiar with the line equation as
> $$y = mx + c$$

8.2 Nature of given data

It is always advisable to think about the type of data involved before any regression analysis is started.

For example:

	x	y
1	Height of mother	Height of daughter at age 21
2	Load carried by lorry	Fuel consumption of lorry
3	Breadth of skull	Length of skull

> In cases **1** and **2**, x can affect y but y can't affect x. Regression line of y on x is appropriate.

> In case **3**, both x and y are influenced by other factors which are not given – correlation is the best analysis.

> If *y* can, sensibly, be predicted from *x*, then
> *y* is called the **dependent** or **response** variable and
> *x* is called the **independent** or **explanatory** variable.

Consider cases **1** and **2**.

The scatter diagrams involved might look like these.

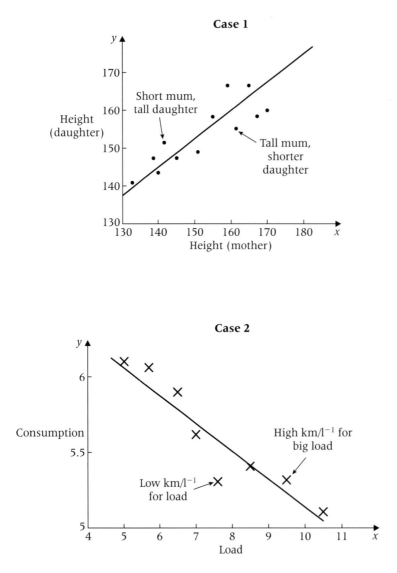

In each case, *x* is the **explanatory** variable and its values are fixed at the start of the experiment.

Clearly, a mother's height is known well before her daughter reaches the age of 21.

The load to be carried by a lorry would be measured before the trial to find fuel consumption.

Some questions to consider

- How can the regression line be obtained?
- How good is the fit of the line?
- What do the positions of the individual points mean?

> The PMCC can be found.
> See comments on the scatter diagrams.

8.3 Residuals

If you return to case **2**, the data supplied was as follows:

x lorry load (000s kg)	5	5.7	6.5	7	7.6	8.5	9.5	10.5
y fuel consumption (km l^{-1})	6.21	6.12	5.90	5.62	5.25	5.41	5.32	5.11

> Note that the fuel consumption is given here in km l^{-1}. A low value indicates that the lorry is using a lot of fuel, while a high value indicates economical fuel usage.

The product moment correlation coefficient (PMCC)

$r = -0.921$ which indicates a strong negative correlation between fuel consumption and load. The points are all close to a straight line.

How close are they and where should the line be placed?

The vertical distances drawn on the scatter diagram are labelled d_i. These are called the **residuals**.

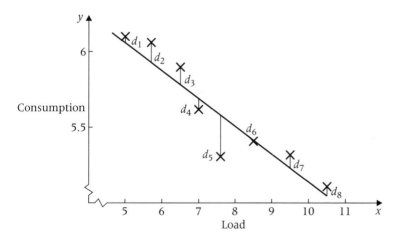

> Residuals can be positive (points **above** line) like d_2, d_3, d_7 or negative (points **below** line) like d_4, d_5.

> Occasionally, a point might lie exactly on the line.

The residuals measure how far away, for the y values, each point is from the line of best fit. The sum of the squares of these distances is minimised to find the line of regression of y on x.

> For each point, the residual is the difference between the observed value of y and the value of y predicted by the line.

8.4 Finding the regression line

The regression line is obtained using the method of least squares and the line is often called the **least squares regression line**.

$d_1^2 + d_2^2 + d_3^2 + \ldots + d_8^2$ or Σd_i^2 is minimised.

The formula is:

$$(y - \bar{y}) = \frac{S_{xy}}{S_{xx}}(x - \bar{x})$$

From this, you should see that the point $(\bar{x}, \bar{y})$ **always** lies on the regression line.

Written in full:

$$(y - \bar{y}) = \frac{\dfrac{\Sigma xy}{n} - \bar{x}\bar{y}}{\dfrac{\Sigma x^2}{n} - \bar{x}^2}(x - \bar{x})$$

The regression equation can be obtained directly using a calculator. This is quite acceptable in an exam.

Using the formulae

For these data (section 8.3)

$$\bar{x} = 7.5375, \qquad \bar{y} = 5.6175, \qquad n = 8,$$

$$\Sigma xy = 333.704, \quad \Sigma x^2 = 479.25.$$

So the equation of regression of y on x using the method of least squares is:

$$(y - 5.6175) = \frac{\left(\dfrac{1}{8} \times 333.704 - 7.5375 \times 5.6175\right)}{\left(\dfrac{1}{8} \times 479.25 - 7.5375^2\right)} \times (x - 7.5375)$$

Remember that this is an equation connecting x and y. y will remain on the left-hand side and x on the right.

So $y - 5.6175 = -0.203\,38\,(x - 7.5375)$

Be very careful not to round prematurely.

and $y = \mathbf{7.15 - 0.203}x$ is the regression equation.

intercept on y-axis   gradient of line
at $x = 0$

The equation can be obtained directly from a calculator. Check carefully as some calculators give the equation as $y = ax + b$ rather than $y = a + bx$.

8.5 Interpretation of line

The regression line gives some important information about the exact nature of the relationship between x and y.

The gradient and intercept values should always be commented upon by reference to the data involved. In this case:

$$a = \mathbf{7.15}$$

In this case, comments are required referring to fuel consumption and load.

This intercept value gives an estimate of the amount of fuel consumption, y, when the load, x, is zero. This tells you that the fuel consumption of an unladen lorry is 7.15 km l^{-1}.

$$b = -0.203$$

7.15 and 0.203 are estimates and are unlikely to be exactly correct.

The gradient indicates, in this case, that y decreases as x increases. Specifically, the fuel consumption, y, decreases by 0.203 km l^{-1} for every extra 1000 kg of load.

8.6 Plotting the regression line

As seen earlier, the point $(\bar{x}, \bar{y})$ always lies on the least squares regression line. This point should be plotted on your scatter diagram.

$(\bar{x}, \bar{y}) \approx (7.54, 5.62)$

To complete plotting the line accurately, one or preferably two other points should be plotted.

Warning! Never assume that any of the given data points will lie on the line.

Any suitable values for x can be chosen but they need to be spread out over the given range. For example;

when $x = 5.5$ gives $\hat{y} = 6.0$ $(7.15 - 0.203 \times 5.5)$

and when $x = 10$ gives $\hat{y} = 5.1$ $(7.15 - 0.203 \times 10)$.

$\hat{y}$ means an *estimated* value of y. Many calculators will find $\hat{y}$ directly for a given x.

The three points (5.5, 6.0) (10, 5.1) and (7.54, 5.62) can be joined to draw the regression line.

8

8.7 Further use of residuals

Consider the data we are examining in case **2** but imagine that more information has now become available. It has been discovered that three different drivers, Ahmed (A), Brian (B) and Carole (C), were involved in the trial. Their individual results were:

Driver	C	B	C	A	A	C	B	B
x load (000s kg)	5	5.7	6.5	7	7.6	8.5	9.5	10.5
y consumption (km l^{-1})	6.21	6.12	5.90	5.62	5.25	5.41	5.32	5.11

The scatter diagram can now be further labelled with this information.

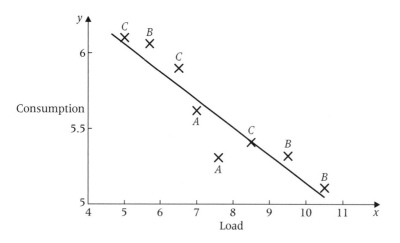

Considering the residual values, further deductions can be made.

It seems that Carole's fuel consumption values lie close to the predicted values given by the line of best fit. However, Ahmed achieves fuel consumption **well below** those predicted by the line and Brian achieves **high** fuel consumption.

> **Note:**
> - Use common sense in your interpretations
> - Always refer to the data given.

This extra information is worth commenting on but always be careful not to make rash judgements, as there may well be other factors involved, for example:

Did all drivers have the same model of lorry?

Did all drivers have the same age of lorry?

Were the journeys of similar length?

Were the journeys over similar types of roads?

> You may well think of other factors.

It would be very unfair on Ahmed if it was immediately assumed that he was a 'bad' driver and he was sacked.

8.8 Predictions

How could the transport manager of the freight company that Ahmed, Brian and Carole work for, use the regression line to predict the fuel consumption for the delivery of a specific load?

Several factors need to be considered:

- How close to the line are the points – is the regression line a 'good fit'?

> PMCC measures this, see chapter 7.

- Is it sensible to predict *y* from *x*?

> Look back to section 4.3.

- What is the range of the *x*-values given from which the line was calculated?

> These are the *x*-values given in the table of results.

- What is the size of the *x*-value from which a value of *y* is to be predicted?

For any regression line where the fit is good, and it is sensible to predict y from x, a value of y can be obtained by substituting a value of x into $y = a + bx$.

There are **limitations** to these predictions however.

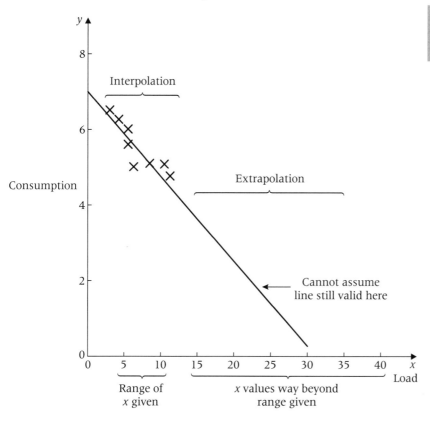

Look at this scatter diagram where the scale on the x-axis has been extended.

Interpolation

It is perfectly valid to use the least squares regression line $y = 7.15 - 0.203x$ to predict fuel consumption for loads between 5000 and 10 500 kg.

$5 \leq x \leq 10.5$ is the given range.

This is called **interpolation**.

For example, to find the predicted fuel consumption for a load of 8000 kg.

8000 kg means $x = 8$.

$$\hat{y} = 7.15 - 0.203 \times 8 = 5.53 \text{ km l}^{-1} \text{ (to 3 s.f.)}.$$

$\hat{y}$ is the notation for an *estimated* value of y, see section 8.6.

Extrapolation

Could the fuel consumption be predicted if the load was 30 000 kg ($x = 30$)?

Clearly, common sense would indicate that this load is enormous compared to those loads given in the original data. The lorry would probably collapse under the weight!

? Fuel consumption

8000 kg

Predicting *y* from *x*, when *x* is outside the range of given data is called **extrapolation** and is very dangerous as the *y*-value obtained is likely to be completely inaccurate.

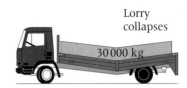

Lorry collapses

Examples

This scatter diagram shows the time in seconds to run 100 m against the number of weeks of intensive training undertaken by the athlete.

There appears to be a linear relationship but clearly this cannot continue indefinitely. The number of seconds taken to run 100 m will not keep on decreasing but will probably level off when the athlete is fully fit and trained for the event.

The second scatter diagram illustrates a possible relationship between the amount of fertiliser used and the yield from a plot of tomato plants.

Again, it appears that there is a strong linear relationship between yield and amount of fertiliser, but clearly the yield will not continue to increase in this way and, in fact, it will probably decline as too much fertiliser may well lead to a decrease in tomato yield.

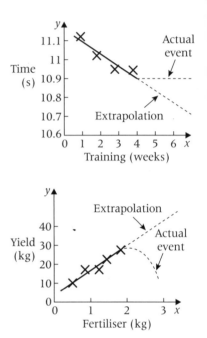

Worked example 8.1

1 An electric fire was turned on in a cold room and the temperature of the room was noted at five-minute intervals.

Time from switching on fire, x (min)	0	5	10	15	20	25	30	35	40
Temperature, y (°C)	0.4	1.5	3.4	5.5	7.7	9.7	11.7	13.5	15.4

(a) Plot the data on a scatter diagram.

(b) Calculate the line of regression $y = a + bx$ and draw it on your scatter diagram.

(c) Predict the temperature 60 minutes from switching on the fire. Why should this prediction be treated with caution?

(d) Explain why, in **(b)** the line $y = a + bx$ was calculated.

(e) If, instead of the temperature being measured at five-minute intervals, the time for the room to reach predetermined temperatures (e.g. 1, 4, 7, 10, 13 °C) had been observed, what would the appropriate calculation have been?

Solution

(a)

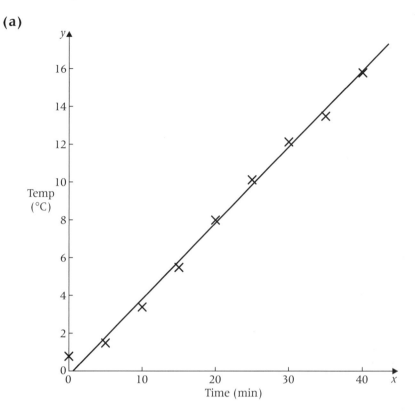

(b) $y = -0.142 + 0.389x$

$\bar{x} = 20, \bar{y} = 7.64$,   plot (20, 7.64)

$x = 10, \hat{y} = 3.75$,   plot (10, 3.75)

$x = 37.5, \hat{y} = 14.46$,   plot (37.5, 14.46)

(c) $x = 60, \hat{y} = 23.2$.

Treat with caution because 60 minutes is outside the range of times given. The linear model cannot continue indefinitely as the room cannot keep on heating up forever.

(d) The line $y = a + bx$, the equation of temperature on time, was used because y depended on the value of x. The dependent variable y was observed at predetermined values of the explanatory variable x.

(e) If the time to reach a temperature was observed, then x would be observed at predetermined values for y. In this case, x would depend on the value of y, so an equation of time on temperature would be appropriate.

> The equation of the line of regression can be obtained directly using a calculator.

> $\hat{y}$ can be obtained using a calculator or by substituting into the regression equation, so for $x = 10$,
> $\hat{y} = -0.142 + 0.389 \times 10 = 3.75$

8

Worked example 8.2

The following data refer to a particular developed country. The table shows for each year, the annual average temperature $x°C$, and an estimate of the total annual domestic energy consumption, y PJ (peta-joules).

Year	x	y
1984	9.6	1664
1985	9.3	1715
1986	9.8	1622
1987	10.3	1624
1988	10.1	1621
1989	10.4	1588
1990	10.8	1577
1991	9.7	1719
1992	10.7	1604
1993	9.2	1811
1994	9.8	1754

(a) Illustrate the relationship between energy consumption and temperature by a scatter diagram. Label the points according to the year.

(b) Calculate the line of regression of y on x and draw the line on your scatter diagram.

(c) Use your equation to estimate the energy consumption in 1995, given that the average temperature in that year was 10.3°C.

(d) Calculate residuals for each of the years 1991, 1992, 1993 and 1994. Comment on their values and interpret the pattern shown by the scatter diagram.

(e) Modify your estimate of energy consumption in 1995 in the light of the residuals you have calculated.

Solution

(a)

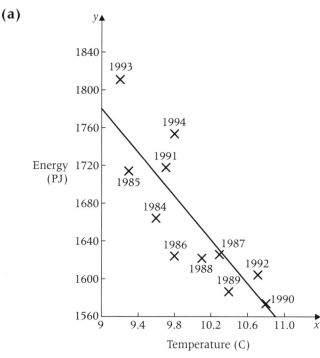

(b) $y = 2849 - 118.9x$

$\bar{x} = 9.97, \bar{y} = 1663.5$, plot $(9.97, 1663.5)$

$x = 9, \hat{y} = 1778.9$, plot $(9, 1778.9)$

$x = 10, \hat{y} = 1660$, plot $(10, 1660)$

> The regression equation can be obtained directly using a calculator.

> Obtain $\hat{y}$ from a calculator or by substituting into the equation as in the previous example.

(c) Prediction for 1995

$x = 10.3 \quad \hat{y} = 1625$

(d) Residuals are obtained in the following way.

> Residuals can be found using the equation or by reading off the vertical difference between the value of y given by the line and the observed value of y.

Year	y	x	$\hat{y} =$ $2849 - 118.9x$	Residual $=$ $y - \hat{y}$
1991	1719	9.7	1695.7	23.3
1992	1604	10.7	1576.8	27.2
1993	1811	9.2	1755.1	55.9
1994	1754	9.8	1683.8	70.2

> Clear, common sense comments are needed.

The residuals seem to be increasing with time. The scatter diagram suggests that energy consumption decreases as the average temperature increases. There is also an increase in energy consumption with time.

(e) The modified estimate for 1995 would be

$$1625 + 85 = 1710.$$

> The residuals found in **(d)** clearly indicate a higher value would be expected in 1995.
> A sensible suggestion would be to add a residual value on to the 1625 prediction. This value will be greater than 70.2 for 1994. Continuing the pattern, approx. $70.2 + 15 = 85$ might be the residual in 1995.

8

EXERCISE 8A

1 The heart and body mass of 14 10-month-old male mice are given in the following table.

Body mass, x (g)	Heart mass, y (mg)
27	118
30	136
37	156
38	150
32	140
36	155
32	157
32	114
38	144
42	159
36	149
44	170
33	131
38	160

(a) Draw a scatter diagram of these data.

(b) Calculate the line of regression of heart mass on body mass (y on x).

2 The systolic blood pressure of ten men of various ages are given in the following table.

Age, x (years)	Systolic blood pressure, y (mm Hg)
37	110
35	117
41	125
43	130
42	138
50	146
49	148
54	150
60	154
65	160

(a) Draw a scatter diagram.

(b) Find the line of regression of systolic blood pressure on age.

(c) Use your line to predict the systolic blood pressure for a man who is:
(i) 20 years old,
(ii) 45 years old.

(d) Comment on the likely accuracy of your predictions in **(i)** and **(ii)**.

3 A scientist, working in an agricultural research station, believes that there is a relationship between the hardness of the shells of the eggs laid by chickens and the amount of a certain food supplement put into the diet of the chickens. He selects ten chickens of the same breed and collects the following data.

Chicken	Amount of supplement, x (g)	Hardness of shell, y
A	7.0	1.2
B	9.8	2.1
C	11.6	3.4
D	17.5	6.1
E	7.6	1.3
F	8.2	1.7
G	12.4	3.4
H	17.5	6.2
I	9.5	2.1
J	19.5	7.1

(Hardness is measured on a scale of 0–10, 10 being the hardest. No units are attached.)

(a) Draw a scatter diagram to illustrate these data.

(b) Calculate the equation of the regression line of hardness on amount of supplement.

(c) Do you believe that this linear model will continue to be appropriate no matter how large or small x becomes? Justify your reply.

4 In an investigation into predictions using the stars, a well-known astrologer, Horace Scope, predicted the ages at which 13 young people would first marry. The completed data, of predicted and actual ages at first marriage, are now available and are summarised in the following table.

Person	Predicted age, x (years)	Actual age, y (years)
A	24	23
B	30	31
C	28	28
D	36	35
E	20	20
F	22	25
G	31	45
H	28	30
I	21	22
J	29	27
K	40	40
L	25	27
M	27	26

8

(a) Draw a scatter diagram of these data.

(b) Calculate the line of regression of y on x.

(c) Plot the regression line on the scatter diagram.

(d) Comment on the results obtained, particularly in view of the data for person G. What further action would you suggest?　　　　　　　　　　　　　　　　[A]

5 The given data relate to the price and engine capacity of new cars in January 1982.

Car model	Price (£) y	Capacity (cc) x
A	3900	1000
B	4200	1270
C	5160	1750
D	6980	2230
E	6930	1990
F	2190	600
G	2190	650
H	4160	1500
J	3050	1450
K	6150	1650

(a) Plot a scatter diagram of the data.

(b) Calculate the line of regression of y on x.

(c) Draw the line of regression on the scatter diagram.

(d) A particular customer regards large engine capacity and a low price as the two most important factors in choosing a car. Examine your scatter diagram and the regression line to suggest to him one model which, in January 1982, gave good value for money. Also suggest three models which you would advise the customer not to buy.　　　　　　　　　　　　[A]

6 A small firm tries a new approach to negotiating the annual pay rise with each of its 12 employees. In an attempt to simplify the process, it is suggested that each employee should be assigned a score, x, based on his/her level of responsibility. The annual salary will be £$(a + bx)$ and negotiations will only involve the values of a and b.

The following table gives last year's salaries (which were generally regarded as fair) and the proposed scores.

Employee	x	Annual salary (£) y
A	10	5750
B	55	17 300
C	46	14 750
D	27	8200
E	17	6350
F	12	6150
G	85	18 800
H	64	14 850
I	36	9900
J	40	11 000
K	30	9150
L	37	10 400

8

(a) Plot the data on a scatter diagram.

(b) Estimate the values that could have been used for a and b last year by finding the line of regression of y on x.

(c) Comment on whether the suggested method is likely to prove reasonably satisfactory in practice.

(d) Two employees, B and C, had to work away from home for a large part of the year. In the light of this additional information, suggest an improvement to the model.   [A]

7 A company specialises in supplying 'stocking fillers' at Christmas time. The company employs several full-time workers all year round but it relies on part-time help at the Christmas rush period.

The time taken to pack orders, together with the packer concerned and the number of items in the order were recorded for orders chosen at random during three days just prior to Christmas.

Packer	No. of items x	Time to pack y (mins)
Ada	21	270
Ada	62	420
Betty	30	245
Alice	20	305
Ada	35	320
Ada	57	440
Alice	40	400
Betty	10	180
Ada	48	350
Alice	58	490
Ada	20	285
Betty	45	340

(a) Plot a scatter diagram to illustrate these data. Label clearly which packer was responsible for the order.

(b) Calculate the value of the product moment correlation coefficient and comment on its value.

(c) Find the line of regression of y on x.

(d) Find an estimate for the length of time that it would take to pack an order of 45 items. Comment on how good an estimate you would imagine this to be.

(e) Calculate the residual values for Betty and also for her daughter Alice, who is working in the factory on a temporary basis over the Christmas holiday.

(f) Use these residuals to produce a better estimate of the actual time expected for the next order of 45 items to be assembled if:

  (i) Betty is the packer,

  (ii) Alice is the packer.

8 Over a period of three years, a company has been monitoring the number of units of output produced per quarter and the total cost of producing the units. The table below shows the results.

Units of output, x (1000s)	Total cost, y (£1000)
14	35
29	50
55	73
74	93
11	31
23	42
47	65
69	86
18	38
36	54
61	81
79	96

(a) Draw a scatter diagram of these data.

(b) Calculate the equation of the regression line of y on x and draw this line on your scatter diagram.

The selling price of each unit of output is £1.60.

(c) Use your graph to estimate the level of output at which the total income and the total costs are equal.

(d) Give a brief interpretation of this value. [A]

9 In the development of a new plastic material, a variable of interest was its 'deflection' when subjected to a constant force underwater. It was believed that, over a limited range of temperatures, this would be approximately linearly related to the temperature of the water. The 'deflection' was measured at a series of predetermined temperatures with the following results.

Technician	'Deflection' y	Temperature $x\,(°C)$
A	2.05	15
B	2.45	20
A	2.50	25
C	2.00	30
B	3.25	35
A	3.20	40
C	4.50	45
B	3.85	50
A	3.70	55
C	3.65	60

(a) Illustrate this data with a scatter diagram.

(b) Calculate the equation of the regression line of 'deflection' on temperature and draw this line on your scatter diagram.

(c) Three different technicians, A, B and C, were involved in the trial. Label your scatter diagram with this information and comment on the performance of each technician.

(d) Suggest what action might be taken before conducting further trials. [A]

10 In addition to its full-time staff, a supermarket employs part-time sales staff on Saturdays. The manager experimented to see if there is a relationship between the takings and the number of part-time staff employed.

He collected data over nine successive Saturdays.

Number of part-time staff employed, x	Takings, £'00 y
10	313
13	320
16	319
19	326
22	333
25	342
28	321
31	361
34	355

(a) Plot a scatter diagram of these data.

(b) Calculate the equation of the regression line of takings on the number of part-time staff employed. Draw the line on your scatter diagram.

(c) If the regression line is denoted by $y = a + bx$, give an interpretation to each of a and b.

(d) On one Saturday, major roadworks blocked a nearby road. Which Saturday do you think this was? Give a reason for your choice.

(e) The manager had increased the number of part-time staff each week. This was desirable from an organisational point of view but undesirable from a statistical point of view. Comment. [A]

11 As part of an investigation into how accurately witnesses are able to describe incidents, Paulo observed a number of staged incidents. Afterwards he was asked to estimate the ages of some of the participants. Paulo's estimates, y years, together with the actual ages, x years, are shown in the table below.

Participant	Actual age, x (years)	Estimated age, y (years)
A	86	74
B	55	51
C	28	22
D	69	59
E	45	38
F	7	8
G	17	15
H	11	9
I	37	38
J	2	2
K	78	66

(a) Draw a scatter diagram of Paulo's estimate, y, and the actual age, x.

(b) Calculate the equation of the regression line of y on x and draw it on your diagram.

(c) Calculate the residuals for participants D and I.

(d) Discuss whether a small residual indicates a good estimate of age. Illustrate your answer by making reference to the residuals you have calculated in (c). [A]

12 A pop group undertakes a country-wide tour. It is accompanied by two roadies who pack up the equipment and load it into a removal van after each performance. They also hire local people at each venue to assist with this task. In order to estimate the best number of local people to hire they experiment by varying the number hired, x, and recording the time, y minutes, taken to load the van. The results are tabulated below.

Performance	x	y
1	2	384
2	3	359
3	4	347
4	6	322
5	6	315
6	7	312
7	9	299
8	9	294
9	10	283

8

(a) Plot a scatter diagram of the data.

(b) Calculate the equation of the regression line of y on x and draw the line on your scatter diagram.

(c) If the line is of the form $y = a + bx$ give an interpretation to each of a and b in the context of this question.

(d) Give two reasons why it would be unwise to use the regression equation to estimate the time taken if 50 people were hired.

(e) The data given for the first nine performances shows that at each performance the number of local people hired was either greater than or equal to the number of people hired at the previous performance. Why does this fact make it more difficult to interpret the results?　[A]

Key point summary

1 A scatter diagram should be drawn to judge whether   *p146* linear regression analysis is a sensible option.

2 The nature of the data should be considered to   *p146* determine which is the *independent* or *explanatory* variable (x) and which is the *dependent* or *response* variable (y).

3 The regression line is found using the *method of*   *p148* *least squares* in the form

$$y = \mathbf{a} + \mathbf{b}x$$

This is the regression line of y on x and may be used to predict a value for y from a given value of x.

The equations can be found directly using a calculator with a linear regression mode.

Be careful to note the form in which your calculator presents the equation – it may be as $y = \mathbf{a}x + \mathbf{b}$.

4 Using $y = a + bx$

a estimates the value of y when x is zero.   *pp148, 149*
b estimates the rate of change of y with x.

5 Be very careful when predicting from your line.   *p152* Watch out for extrapolation when predictions can be wildly inaccurate.

Look back to section 8.8.

Never assume a linear model will keep on going forever.

Test yourself		**What to review**

1 For each of the following sets of data say which variable is the response or dependent variable, and which is the explanatory or independent variable.

Section 8.2

(a)

Temperature required, w (°C)	Time taken to reach required temp, u (min)
15	4.3
20	8.7
25	11.9
30	14.8
35	17.1

(b)

Time fire has been switched on, f (min)	Temperature reached g (°C)
5	12.2
10	14.6
15	16.1
20	17.8
25	19.3

2 Which of the following scatter diagrams could illustrate data connected by the given regression equations?

Section 8.4

(a) $y = -6x + 12.3$

(b) $p = 0.78t - 2.1$

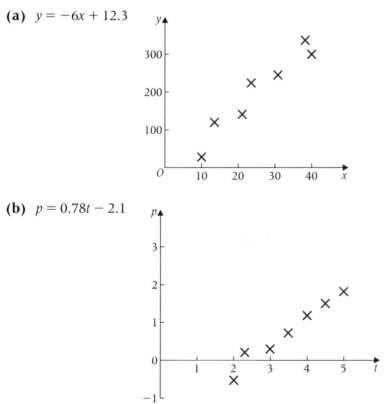

Test yourself (continued)	**What to review**

(c) $m = 0.15b - 1.9$

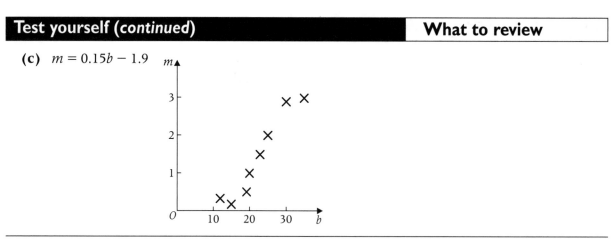

---

3 The line of regression of mass (y kg) on age (x weeks) for baby giraffes, between 0 and 12 weeks of age, is given below:

Sections 8.2 and 8.5

$$y = 2.07x + 21.7$$

(a) Is it possible to obtain an estimate of the mass of a baby giraffe at 9 weeks old from this equation?

(b) Is it possible to estimate the age in weeks of a baby giraffe which has mass 42 kg?

(c) Interpret the value of the constant 21.7 in this equation.

---

4 The table below gives the height of a bean shoot in centimetres (y) and the number of days since it was planted (x).

Sections 8.4 and 8.8

Number of days, x	Height, y (cm)
40	9.6
45	10.5
50	11.2
55	12.3
60	13.4
65	14.3
70	15.2

(a) Calculate the line of regression of y on x.

(b) Estimate the height of the shoot exactly eight weeks (56 days) after planting.

(c) Why would it not be sensible to use the regression equation to estimate the height of the shoot three months after planting?

Test yourself (continued)		**What to review**

5 As part of his research into the behaviour of the human memory, a leading psychologist asked 15 schoolgirls from years 9, 10 and 11 to talk for five minutes on 'my day at school'. The psychologist asked each girl to record how many times she thought she had used the word 'nice' during the talk. The following table gives their replies together with the true values.

Section 8.3

Girl	True value x	Recorded value y
A	12	9
B	20	19
C	1	3
D	8	14
E	0	4
F	12	12
G	12	16
H	17	14
I	6	5
J	5	9
K	24	20
L	23	16
M	10	11
N	18	17
O	16	19

The equation of the regression line of y on x is $y = 4.40 + 0.663x$.

The girls are from three different year groups.

A, *C*, *H*, *I* and *L* are from Year 11.

E, *F*, *K*, *M* and *N* are from Year 10.

B, *D*, *G*, *J* and *O* are from Year 9.

Find the residuals for the girls in Year 9 and for those in Year 11. Use these, together with the regression line to estimate:

(a) the recorded value for a girl in Year 9 whose true value was 15,

(b) the recorded value for a girl in Year 11 whose true value was 10.

8

1 (a) Explanatory: temperature, dependent: time

 (b) Explanatory: time, dependent: temperature

2 (a) No; regression equation implies a negative gradient

 (b) Yes;

 (c) Yes;

3 (a) Yes;

 (b) No (line has y as dependent not x);

 (c) Mass at birth: 21.7 kg.

4 (a) $y = 0.190x + 1.91$;

 (b) $y = 0.190 \times 56 + 1.91 = 12.5$ cm;

 (c) At 3 months, extrapolation would be used and therefore results may be very inaccurate as linear model may not continue.

5 Residuals: B 1.3, D 4.3, G 3.6, J 1.3, O 4.0,
    A −3.4, C −2.1, H −1.7, I −3.4, L −3.7.
    (a) 17; (b) 8.

Exam style practice paper

Time allowed 1 hour 30 minutes – Candidates taking the course work option will sit a slightly shorter paper (1 hour 15 minutes) of the same standard.

Answer **all** questions

---

1 Jeremy sells a magazine which is produced in order to raise money for homeless people. The probability of making a sale is, independently, 0.09 for each person he approaches. Given that he approaches 40 people, find the probability that he will make:

   (a) two or fewer sales, *(3 marks)*

   (b) exactly four sales, *(2 marks)*

   (c) more than five sales. *(2 marks)*

2 Explain the difference between a parameter and a statistic.
   (3 marks)

3 A rugby club has three categories of membership: adult, social and junior. The number of members in each category, classified by gender, is shown in the table.

	Adult	Social	Junior
Female	25	35	40
Male	95	25	80

One member is chosen, at random, to cut the ribbon at the opening of the new clubhouse.
V denotes the event that a female member is chosen.
W denotes the event that an adult member is chosen.
X denotes the event that a junior member is chosen.

   (a) Find:

      (i) $P(V)$,

      (ii) $P(X)$,

      (ii) $P(X \mid V)$. *(4 marks)*

   (b) For the events V, W and X, write down two which are:

      (i) mutually exclusive (justify your answer), *(2 marks)*

      (ii) independent (justify your answer). *(2 marks)*

4 A small firm wishes to introduce an aptitude test for applicants for assembly work. The test consists of a mechanical puzzle. The assembly workers, currently employed, were asked to complete the puzzle. They were timed to the nearest second and the times taken by 35 of them are shown below.

Time to complete puzzle (s)	Frequency
20–39	6
40–49	8
50–54	7
55–59	5
60–99	9

(a) Estimate the median and the interquartile range of the data. *(5 marks)*

(b) Calculate estimates of the mean and the standard deviation of the data. *(4 marks)*

(c) In addition to the data in the table, five other assembly workers completed the puzzle but took so long to do so that their times were not recorded. These times all exceeded 100 s. Estimate the median time to complete the puzzle for all 40 assembly workers. *(2 marks)*

(d) The firm decides not to offer employment to any applicant who takes longer to complete the puzzle than the average time taken by the assembly workers who took the test.

    (i) State whether you would recommend the median or the mean to be used as a measure of average in these circumstances. Explain your answer. *(2 marks)*

    (ii) Write down the value of your recommended measure of average. *(1 mark)*

5 Applicants to join a police force are tested for physical fitness. Based on their performance, a physical fitness score is calculated for each applicant. Assume that the distribution of scores is normal.

(a) The scores for a random sample of ten applicants were

     55   23   44   69   22   45   54   72   34   66

Experience suggests that the standard deviation of scores is 14.8. Calculate a 99% confidence interval for the mean score of all applicants. *(5 marks)*

(b) The scores of a further random sample of 110 applicants had a mean of 49.5 and a standard deviation of 16.5. Use the data from this second sample to calculate:

   (i) a 95% confidence interval for the mean score of all
        applicants, *(3 marks)*
   (ii) an interval within which the score of approximately
        95% of applicants will lie. *(2 marks)*
(c) By interpreting your results in **(b)(i)** and **(b)(ii)**,
      comment on the ability of applicants to achieve a score
      of 25. *(3 marks)*

6 [A sheet of graph paper is provided for use in this question.]
   Hennie, a statistics teacher, is an unenthusiastic gardener.
   During the summer months she records the time, y minutes,
   it takes her to cut her lawn together with the time, x days,
   since she last cut it.

x	5	13	10	19	7	12	21	18	24	12
y	23	32	28	48	21	35	44	39	50	39

(a) Plot a scatter diagram of the data. *(3 marks)*
(b) Find the equation of the regression line of y on x and
      plot it on your scatter diagram. *(6 marks)*
(c) Give an interpretation, in the context of this questions,
      of:
   (i) the gradient of the regression line,
   (ii) the intercept of the regression line with the y-axis.
        (3 marks)
(d) Hennie's brother, Ludwig, suggests to her that she could
      save time by cutting her lawn in spring and then waiting
      until the autumn before cutting it again. Give one
      statistical reason, and one reason specific to this context,
      why it would be unwise to use your regression equation
      to estimate the time it would take Hennie to cut her
      lawn 150 days since she last cut it. *(2 marks)*
(e) Explain why it was appropriate to this question to
      calculate the regression equation of y on x rather than
      that of x on y. *(2 marks)*

7 Rahul works for a bank. He can travel to the branch where he
   works by car, bus or bicycle. He does not wish to arrive too
   early as he cannot enter the building before it is unlocked at
   8.30 a.m. He does not wish to arrive after his starting time,
   which is 9.00 a.m.

(a) When he travels by bus his journey time may be
      modelled by a normally distributed random variable with
      mean 40 minutes and standard deviation 10 minutes.
      Given that he leaves home at 8.00 a.m. and travels by
      bus, find the probability that he will arrive at the branch:
   (i) before 9.00 a.m.
   (ii) between 8.30 a.m. and 9.00 a.m. *(5 marks)*

(b) When he travels by bicycle his journey time may be modelled by a normally distributed random variable with mean 35 minutes and standard deviation 2 minutes. What time should he leave home when he travels by bicycle if he wishes to have a probability of 0.99 of arriving before 9.00 a.m.? *(5 marks)*

(c) When he travels by car his journey time may be modelled by a normally distributed random variable with mean 28 minutes and standard deviation 8 minutes. What time should he leave home when he travels by car, in order to maximise his chance of arriving between 8.30 a.m. and 9.00 a.m. *(2 marks)*

(d) Give an advantage of:
    (i) travelling by car compared to bus or bicycle,
                                        (1 mark)
    (ii) travelling by bicycle compared to bus or car.
                                        (1 mark)

Your answers to **(d)** should be based only on the data given in the question and not on general considerations such as exercise, environmental consequences and comfort.

Appendix

Table 1 Binomial distribution function

The tabulated value is $P(R \leqslant r)$, where R has a binomial distribution with parameters n and p.

r	0.01	0.02	0.03	0.04	0.05	0.06	0.07	0.08	0.09	0.10	0.15	0.20	0.25	0.30	0.35	0.40	0.45	0.50	r
n = 8 0	0.9227	0.8508	0.7837	0.7214	0.6634	0.6096	0.5596	0.5132	0.4703	0.4305	0.2725	0.1678	0.1001	0.0576	0.0319	0.0168	0.0084	0.0039	0
1	0.9973	0.9897	0.9777	0.9619	0.9428	0.9208	0.8965	0.8702	0.8423	0.8131	0.6572	0.5033	0.3671	0.2553	0.1691	0.1064	0.0632	0.0352	1
2	0.9999	0.9996	0.9987	0.9969	0.9942	0.9904	0.9853	0.9789	0.9711	0.9619	0.8948	0.7969	0.6785	0.5518	0.4278	0.3154	0.2201	0.1445	2
3	1.000	1.000	0.9999	0.9998	0.9996	0.9993	0.9987	0.9978	0.9966	0.9950	0.9786	0.9437	0.8862	0.8059	0.7064	0.5941	0.4770	0.3633	3
4			1.000	1.000	1.000	1.000	0.9999	0.9999	0.9997	0.9996	0.9971	0.9896	0.9727	0.9420	0.8939	0.8263	0.7396	0.6367	4
5							1.000	1.000	1.000	1.000	0.9998	0.9988	0.9958	0.9887	0.9747	0.9502	0.9115	0.8555	5
6											1.000	0.9999	0.9996	0.9987	0.9964	0.9915	0.9819	0.9648	6
7												1.000	1.000	0.9999	0.9998	0.9993	0.9983	0.9961	7
8														1.000	1.000	1.000	1.000	1.000	8
n = 12 0	0.8864	0.7847	0.6938	0.6127	0.5404	0.4759	0.4186	0.3677	0.3225	0.2824	0.1422	0.0687	0.0317	0.0138	0.0057	0.0022	0.0008	0.0002	0
1	0.9938	0.9769	0.9514	0.9191	0.8816	0.8405	0.7967	0.7513	0.7052	0.6590	0.4435	0.2749	0.1584	0.0850	0.0424	0.0196	0.0083	0.0032	1
2	0.9998	0.9985	0.9952	0.9893	0.9804	0.9684	0.9532	0.9348	0.9134	0.8891	0.7358	0.5583	0.3907	0.2528	0.1513	0.0834	0.0421	0.0193	2
3	1.000	0.9999	0.9997	0.9990	0.9978	0.9957	0.9925	0.9880	0.9820	0.9744	0.9078	0.7946	0.6488	0.4925	0.3467	0.2253	0.1345	0.0730	3
4		1.000	1.000	0.9999	0.9998	0.9996	0.9991	0.9984	0.9973	0.9957	0.9761	0.9274	0.8424	0.7237	0.5833	0.4382	0.3044	0.1938	4
5				1.000	1.000	1.000	0.9999	0.9998	0.9997	0.9995	0.994	0.9806	0.9456	0.8822	0.7873	0.6652	0.5269	0.3872	5
6							1.000	1.000	1.000	0.9999	0.9993	0.9961	0.9857	0.9614	0.9154	0.8418	0.7393	0.6128	6
7										1.000	0.999	0.994	0.9972	0.9905	0.9745	0.9427	0.8883	0.8062	7
8											1.000	0.9999	0.9996	0.9983	0.9944	0.9847	0.9644	0.9270	8
9												1.000	1.000	0.9998	0.9992	0.9972	0.9921	0.9807	9
10														1.000	0.9999	0.9997	0.9989	0.9968	10
11															1.000	1.000	0.9999	0.9998	11
12																	1.000	1.000	12
n = 15 0	0.8601	0.7386	0.6333	0.5421	0.4633	0.3953	0.3367	0.2863	0.2430	0.2059	0.0874	0.0352	0.0134	0.0047	0.0016	0.0005	0.0001	0.0000	0
1	0.9904	0.9647	0.9270	0.8809	0.8290	0.7738	0.7168	0.6597	0.6035	0.5490	0.3186	0.1671	0.0802	0.0353	0.0142	0.0052	0.0017	0.0005	1
2	0.9996	0.9970	0.9906	0.9797	0.9638	0.9429	0.9171	0.8870	0.8531	0.8159	0.6042	0.3980	0.3980	0.2361	0.1268	0.0271	0.0107	0.0037	2
3	1.000	0.9998	0.9992	0.9976	0.9945	0.9896	0.9825	0.9727	0.9601	0.9444	0.8227	0.6482	0.4613	0.2969	0.1727	0.0905	0.0424	0.0176	3
4		1.000	0.9999	0.9998	0.9994	0.9986	0.9972	0.9950	0.9918	0.9873	0.9383	0.8358	0.6865	0.5155	0.3519	0.2173	0.1204	0.0592	4
5			1.000	1.000	0.9999	0.9999	0.9997	0.9993	0.9987	0.9978	0.9832	0.9389	0.8516	0.7216	0.5643	0.4032	0.2608	0.1509	5
6					1.000	1.000	1.000	0.9999	0.9998	0.9997	0.9964	0.9819	0.9434	0.8689	0.7548	0.6098	0.4522	0.3036	6
7								1.000	1.000	1.000	0.9994	0.9958	0.9827	0.9500	0.8868	0.7869	0.6535	0.5000	7
8											0.9999	0.9992	0.9958	0.9848	0.9578	0.9050	0.8182	0.6964	8
9											1.000	0.9999	0.9992	0.9963	0.9876	0.9662	0.9231	0.8491	9
10												1.000	0.9999	0.9993	0.9972	0.9907	0.9745	0.9408	10
11													1.000	0.9999	0.9995	0.9981	0.9937	0.9824	11
12														1.000	0.9999	0.9997	0.9989	0.9963	12
13															1.000	1.000	0.9999	0.9995	13
14																	1.000	1.000	14

Note: the AQA formulae book contains more values of n than are given here.

Table 1 Binomial distribution function (continued)

n = 20

r	0.01	0.02	0.03	0.04	0.05	0.06	0.07	0.08	0.09	0.10	0.15	0.20	0.25	0.30	0.35	0.40	0.45	0.50	r
0	0.8179	0.6676	0.5438	0.4420	0.3585	0.2901	0.2342	0.1887	0.1516	0.1216	0.0388	0.0115	0.0032	0.0008	0.0002	0.0000	0.0000	0.0000	0
1	0.9831	0.9401	0.8802	0.8103	0.7358	0.6605	0.5869	0.5169	0.4516	0.3917	0.1756	0.0692	0.0243	0.0076	0.0021	0.0005	0.0001	0.0000	1
2	0.9990	0.9929	0.9790	0.9561	0.9245	0.8850	0.8390	0.7879	0.7334	0.6769	0.4049	0.2061	0.0913	0.0355	0.0121	0.0036	0.0009	0.0002	2
3	1.000	0.9994	0.9973	0.9926	0.9841	0.9710	0.9529	0.9294	0.9007	0.8670	0.6477	0.4114	0.2252	0.1071	0.0444	0.0160	0.0049	0.0013	3
4		1.000	0.9997	0.9990	0.9974	0.9944	0.9893	0.9817	0.9710	0.9568	0.8298	0.6296	0.4148	0.2375	0.1182	0.0510	0.0189	0.0059	4
5			1.000	0.9999	0.9997	0.9991	0.9981	0.9962	0.9932	0.9887	0.9327	0.8042	0.6172	0.4164	0.2454	0.1256	0.0553	0.0207	5
6				1.000	1.000	0.9999	0.9997	0.9994	0.9987	0.9976	0.9781	0.9133	0.7858	0.6080	0.4166	0.2500	0.1299	0.0577	6
7						1.000	1.000	0.9999	0.9998	0.9996	0.9941	0.9679	0.8982	0.7723	0.6010	0.4159	0.2520	0.1316	7
8								1.000	1.000	0.9999	0.9987	0.9900	0.9591	0.8867	0.7624	0.5956	0.4143	0.2517	8
9										1.000	0.9998	0.9974	0.9861	0.9520	0.8782	0.7553	0.5914	0.4119	9
10											1.000	0.9994	0.9961	0.9829	0.9468	0.8725	0.7507	0.5881	10
11												0.9999	0.9991	0.9949	0.9804	0.9435	0.8692	0.7483	11
12												1.000	0.9998	0.9987	0.9940	0.9790	0.9420	0.8684	12
13													1.000	0.9997	0.9985	0.9935	0.9786	0.9423	13
14														1.000	0.9997	0.9984	0.9936	0.9793	14
15															1.000	0.9997	0.9985	0.9941	15
16																1.000	0.9997	0.9987	16
17																	1.000	0.9998	17
18																		1.000	18

n = 25

r	0.01	0.02	0.03	0.04	0.05	0.06	0.07	0.08	0.09	0.10	0.15	0.20	0.25	0.30	0.35	0.40	0.45	0.50	r
0	0.7778	0.6035	0.4670	0.3604	0.2774	0.2129	0.1630	0.1244	0.0946	0.0718	0.0172	0.0038	0.0008	0.0001	0.0000	0.0000	0.0000	0.0000	0
1	0.9742	0.9114	0.8280	0.7358	0.6424	0.5527	0.4696	0.3947	0.3286	0.2712	0.0931	0.0274	0.0070	0.0016	0.0003	0.0001	0.0000	0.0000	1
2	0.9980	0.9868	0.9620	0.9235	0.8729	0.8129	0.7466	0.6768	0.6063	0.5371	0.2537	0.0982	0.0321	0.0090	0.0021	0.0004	0.0001	0.0000	2
3	0.9999	0.9986	0.9938	0.9835	0.9659	0.9402	0.9064	0.8649	0.8169	0.7636	0.4711	0.2340	0.0962	0.0332	0.0097	0.0024	0.0005	0.0001	3
4	1.000	0.9999	0.9992	0.9972	0.9928	0.9850	0.9726	0.9549	0.9314	0.9020	0.6821	0.4207	0.2137	0.0905	0.0320	0.0095	0.0023	0.0005	4
5		1.000	0.9999	0.9996	0.9988	0.9969	0.9935	0.9877	0.9790	0.9666	0.8385	0.6167	0.3783	0.1935	0.0826	0.0294	0.0086	0.0020	5
6			1.000	1.000	0.9998	0.9995	0.9987	0.9972	0.9946	0.9905	0.9305	0.7800	0.5611	0.3407	0.1734	0.0736	0.0258	0.0073	6
7				1.000	1.000	0.9999	0.9998	0.9995	0.9989	0.9977	0.9745	0.8909	0.7265	0.5118	0.3061	0.1536	0.0639	0.0216	7
8						1.000	1.000	0.9999	0.9998	0.9995	0.9920	0.9532	0.8506	0.6769	0.4668	0.2735	0.1340	0.0539	8
9								1.000	1.000	0.9999	0.9979	0.9827	0.9287	0.8106	0.6303	0.4246	0.2424	0.1148	9
10										1.000	0.9995	0.9944	0.9703	0.9022	0.7712	0.5858	0.3843	0.2122	10
11											0.9999	0.9985	0.9893	0.9558	0.8476	0.7323	0.5426	0.3450	11
12											1.000	0.9996	0.9966	0.9825	0.9396	0.8462	0.6937	0.5000	12
13												0.9999	0.9991	0.9940	0.9745	0.9222	0.8173	0.6550	13
14												1.000	0.9998	0.9982	0.9907	0.9656	0.9040	0.7878	14
15													1.000	0.9995	0.9971	0.9868	0.9560	0.8852	15
16														0.9999	0.9992	0.9957	0.9826	0.9461	16
17														1.000	0.9998	0.9988	0.9942	0.9784	17
18															1.000	0.9997	0.9984	0.9927	18
19																0.9999	0.9996	0.9980	19
20																1.000	0.9999	0.9995	20
21																	1.000	0.9999	21
22																		1.000	22

n = 30

r	0.01	0.02	0.03	0.04	0.05	0.06	0.07	0.08	0.09	0.10	0.15	0.20	0.25	0.30	0.35	0.40	0.45	0.50	r
0	0.7397	0.5455	0.4010	0.2939	0.2146	0.1563	0.1134	0.0820	0.0591	0.0424	0.0076	0.0012	0.0002	0.0000	0.0000	0.0000	0.0000	0.0000	0
1	0.9639	0.8795	0.7731	0.6612	0.5535	0.4555	0.3694	0.2958	0.2343	0.1837	0.0480	0.0105	0.0020	0.0003	0.0000	0.0000	0.0000	0.0000	1
2	0.9967	0.9783	0.9399	0.8831	0.8122	0.7324	0.6487	0.5654	0.4855	0.4114	0.1514	0.0442	0.0106	0.0021	0.0003	0.0000	0.0000	0.0000	2
3	0.9998	0.9971	0.9881	0.9694	0.9392	0.8974	0.8450	0.7842	0.7175	0.6474	0.3217	0.1227	0.0374	0.0093	0.0019	0.0003	0.0000	0.0000	3
4	1.000	0.9997	0.9982	0.9937	0.9844	0.9685	0.9447	0.9126	0.8723	0.8245	0.5245	0.2552	0.0979	0.0302	0.0075	0.0015	0.0002	0.0000	4
5		1.000	0.9998	0.9989	0.9967	0.9921	0.9838	0.9707	0.9519	0.9268	0.7106	0.4275	0.2026	0.0766	0.0233	0.0057	0.0011	0.0002	5
6			1.000	0.9999	0.9994	0.9983	0.9960	0.9918	0.9848	0.9742	0.8474	0.6070	0.3481	0.1595	0.0586	0.0172	0.0040	0.0007	6
7				1.000	0.9999	0.9997	0.9992	0.9980	0.9959	0.9922	0.9302	0.7608	0.5143	0.2814	0.1238	0.0435	0.0121	0.0026	7
8					1.000	1.000	0.9999	0.9996	0.9990	0.9980	0.9722	0.8713	0.6736	0.4315	0.2247	0.0940	0.0312	0.0081	8
9							1.000	0.9999	0.9998	0.9995	0.9903	0.9389	0.8034	0.5888	0.3575	0.1763	0.0694	0.0214	9
10								1.000	1.000	0.9999	0.9971	0.9744	0.8943	0.7304	0.5078	0.2915	0.1350	0.0494	10
11										1.000	0.9992	0.9905	0.9493	0.8407	0.6548	0.4311	0.2327	0.1002	11
12											0.9998	0.9969	0.9784	0.9155	0.7802	0.5785	0.3592	0.1808	12
13											1.000	0.9991	0.9918	0.9599	0.8737	0.7145	0.5025	0.2923	13
14												0.9998	0.9973	0.9831	0.9348	0.8246	0.6448	0.4278	14
15												0.9999	0.9992	0.9936	0.9699	0.9029	0.7691	0.5722	15
16												1.000	0.9998	0.9979	0.9876	0.9519	0.8644	0.7077	16
17													0.9999	0.9994	0.9955	0.9788	0.9286	0.8192	17
18													1.000	0.9998	0.9986	0.9917	0.9666	0.8998	18
19														1.000	0.9996	0.9971	0.9682	0.9506	19
20															0.9999	0.9991	0.9950	0.9786	20
21															1.000	0.9998	0.9984	0.9919	21
22																1.000	0.9996	0.9974	22
23																	0.9999	0.9993	23
24																	1.000	0.9998	24
25																		1.000	25

Table 1 Binomial distribution function (continued)

r	0.01	0.02	0.03	0.04	0.05	0.06	0.07	0.08	0.09	0.10	0.15	0.20	0.25	0.30	0.35	0.40	0.45	0.50	r
n = 40 0	0.6690	0.4457	0.2957	0.1954	0.1285	0.0842	0.0549	0.0356	0.0230	0.0148	0.0015	0.0002	0.0000	0.0000	0.0000	0.0000	0.0000	0.0000	0
1	0.9393	0.8095	0.6615	0.5210	0.3991	0.2990	0.2201	0.1594	0.1140	0.0805	0.0121	0.0015	0.0001	0.0000	0.0000	0.0000	0.0000	0.0000	1
2	0.9925	0.9543	0.8822	0.7855	0.6767	0.5665	0.4625	0.3694	0.2894	0.2228	0.0486	0.0079	0.0010	0.0001	0.0000	0.0000	0.0000	0.0000	2
3	0.9993	0.9918	0.9686	0.9252	0.8619	0.7827	0.6937	0.6007	0.5092	0.4231	0.1302	0.0285	0.0047	0.0006	0.0001	0.0000	0.0000	0.0000	3
4	1.000	0.9988	0.9933	0.9790	0.9520	0.9104	0.8546	0.7868	0.7103	0.6290	0.2633	0.0759	0.0160	0.0026	0.0003	0.0000	0.0000	0.0000	4
5		0.9999	0.9988	0.9951	0.9861	0.9691	0.9419	0.9033	0.8535	0.7937	0.4325	0.1613	0.0433	0.0086	0.0013	0.0001	0.0000	0.0000	5
6		1.000	0.9998	0.9990	0.9966	0.9909	0.9801	0.9624	0.9361	0.9005	0.6067	0.2859	0.0962	0.0238	0.0044	0.0006	0.0001	0.0000	6
7			1.000	0.9998	0.9993	0.9977	0.9942	0.9873	0.9758	0.9581	0.7559	0.4371	0.1820	0.0553	0.0124	0.0021	0.0002	0.0000	7
8				1.000	0.9999	0.9995	0.9985	0.9963	0.9919	0.9845	0.8646	0.5931	0.2998	0.1110	0.0303	0.0061	0.0009	0.0001	8
9					1.000	0.9999	0.9997	0.9990	0.9976	0.9949	0.9328	0.7318	0.4395	0.1959	0.0644	0.0156	0.0027	0.0003	9
10						1.000	0.9999	0.9998	0.9994	0.9985	0.9701	0.8392	0.5839	0.3087	0.1215	0.0352	0.0074	0.0011	10
11							1.000	1.000	0.9999	0.9996	0.9880	0.9125	0.7151	0.4406	0.2053	0.0709	0.0709	0.0032	11
12									1.000	0.9999	0.9957	0.9568	0.8209	0.5772	0.3143	0.1285	0.0386	0.0083	12
13										1.000	0.9986	0.9806	0.8968	0.7032	0.4408	0.2112	0.0751	0.0192	13
14											0.9996	0.9921	0.9456	0.8074	0.5721	0.3174	0.1326	0.0403	14
15											0.9999	0.9971	0.9738	0.8849	0.6946	0.4402	0.2142	0.0769	15
16											1.000	0.9990	0.9884	0.9367	0.7978	0.5681	0.3185	0.1341	16
17												0.9997	0.9953	0.9680	0.8761	0.6885	0.4391	0.2148	17
18												0.9999	0.9983	0.9852	0.9301	0.7911	0.5651	0.3179	18
19												1.000	0.9994	0.9937	0.9637	0.8702	0.6844	0.4373	19
20													0.9998	0.9976	0.9827	0.9256	0.7870	0.5627	20
21													1.000	0.9991	0.9925	0.9608	0.8669	0.6821	21
22														0.9997	0.9970	0.9811	0.9233	0.7852	22
23															0.9989	0.9917	0.9595	0.8659	23
24														1.000	0.9996	0.9966	0.9804	0.9231	24
25															0.9999	0.9988	0.9914	0.9597	25
26															1.000	0.9996	0.9966	0.9808	26
27																0.9999	0.9988	0.9917	27
28																1.000	0.9996	0.9968	28
29																	0.9999	0.9989	29
30																	1.000	0.9997	30
31																		0.9999	31
32																		1.000	32
n = 50 0	0.6050	0.3642	0.2181	0.1299	0.0769	0.0453	0.0266	0.0155	0.0090	0.0052	0.0003	0.0000	0.0000	0.0000	0.0000	0.0000	0.0000	0.0000	0
1	0.9106	0.7358	0.5553	0.4005	0.2794	0.1900	0.1265	0.0827	0.0532	0.0338	0.0029	0.0002	0.0000	0.0000	0.0000	0.0000	0.0000	0.0000	1
2	0.9862	0.9216	0.8108	0.6767	0.5405	0.4162	0.3108	0.2260	0.1605	0.1117	0.0142	0.0013	0.0001	0.0000	0.0000	0.0000	0.0000	0.0000	2
3	0.9984	0.9822	0.9372	0.8609	0.7604	0.6473	0.5327	0.4253	0.3303	0.2503	0.0460	0.0057	0.0005	0.0000	0.0000	0.0000	0.0000	0.0000	3
4	0.9999	0.9968	0.9832	0.9510	0.8964	0.8206	0.7290	0.6290	0.5277	0.4312	0.1121	0.0185	0.0021	0.0002	0.0000	0.0000	0.0000	0.0000	4
5	1.000	0.9995	0.9963	0.9856	0.9622	0.9224	0.8650	0.7919	0.7072	0.6161	0.2194	0.0480	0.0070	0.0007	0.0001	0.0000	0.0000	0.0000	5
6		0.9999	0.9993	0.9964	0.9882	0.9711	0.9417	0.8981	0.8404	0.7702	0.3613	0.1034	0.0194	0.0025	0.0002	0.0000	0.0000	0.0000	6
7		1.000	0.9999	0.9992	0.9968	0.9906	0.9780	0.9562	0.9232	0.8779	0.5188	0.1904	0.0453	0.0073	0.0008	0.0001	0.0000	0.0000	7
8			1.000	0.9999	0.9992	0.9973	0.9927	0.9833	0.9672	0.9421	0.6681	0.3073	0.0916	0.0183	0.0025	0.0002	0.0000	0.0000	8
9				1.000	0.9998	0.9993	0.9978	0.9944	0.9875	0.9755	0.7911	0.4437	0.1637	0.0402	0.0067	0.0008	0.0001	0.0000	9
10					1.000	0.9998	0.9994	0.9983	0.9957	0.9906	0.8801	0.5836	0.2622	0.0789	0.0160	0.0022	0.0002	0.0000	10
11						1.000	0.9999	0.9995	0.9987	0.9968	0.9372	0.7107	0.3816	0.1390	0.0342	0.0057	0.0006	0.0000	11
12							1.000	0.9999	0.9996	0.9990	0.9699	0.8139	0.5110	0.2229	0.661	0.0133	0.0018	0.0002	12
13								1.000	0.9999	0.9997	0.9868	0.8894	0.6370	0.3279	0.1163	0.0280	0.0045	0.0005	13
14									1.000	0.9999	0.9947	0.9393	0.7481	0.4468	0.1878	0.0540	0.0104	0.0013	14
15										1.000	0.9981	0.9692	0.8369	0.5692	0.2801	0.0955	0.220	0.0033	15
16											0.9983	0.9856	0.9017	0.6839	0.3889	0.1561	0.0427	0.0077	16
17											0.9998	0.9937	0.9449	0.7822	0.5060	0.2369	0.0765	0.0164	17
18											0.9999	0.9975	0.9713	0.8594	0.6216	0.3356	0.1273	0.0325	18
19											1.000	0.9991	0.9861	0.9152	0.7264	0.4465	0.12974	0.0595	19
20												0.9997	0.9937	0.9522	0.8139	0.5610	0.2862	0.1013	20
21												0.9999	0.9974	0.9749	0.8813	0.6701	0.3900	0.1611	21
22												1.000	0.9990	0.9877	0.9290	0.7660	0.5019	0.2399	22
23													0.9996	0.9944	0.9604	0.8438	0.6134	0.3359	23
24													0.9999	0.9976	0.9793	0.9022	0.7160	0.4439	24
25													1.000	0.9991	0.9900	0.9427	0.8034	0.5561	25
26														0.9997	0.9955	0.9686	0.8721	0.6641	26
27														0.9999	0.9981	0.9840	0.9220	0.7601	27
28														1.000	0.9993	0.9924	0.9556	0.8389	28
29															0.9997	0.9966	0.9765	0.8987	29
30															0.9999	0.9986	0.9884	0.9405	30
31															1.000	0.9995	0.9947	0.9675	31
32																0.9998	0.9978	0.9836	32
33																0.9999	0.9991	0.9923	33
34																1.000	0.9997	0.9967	34
35																	0.9999	0.9987	35
36																	1.000	0.9995	36
37																		0.9998	37
38																		1.000	38

Table 3 Normal distribution function

The tabulated value is $\Phi(z) = P(Z \leqslant z)$,
where Z is the standardised normal random variable, $N(0, 1)$.

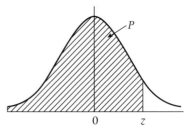

z	0.00	0.01	0.02	0.03	0.04	0.05	0.06	0.07	0.08	0.09	z
0.0	0.50000	0.50399	0.50798	0.51197	0.51595	0.51994	0.52392	0.52790	0.53188	0.53586	0.0
0.1	0.53983	0.54380	0.54776	0.55172	0.55567	0.55962	0.56356	0.56749	0.57142	0.57535	0.1
0.2	0.57926	0.58317	0.58706	0.59095	0.59483	0.59871	0.60257	0.60642	0.61026	0.61409	0.2
0.3	0.61791	0.62172	0.62552	0.62930	0.63307	0.63683	0.64058	0.64431	0.64803	0.65173	0.3
0.4	0.65542	0.65910	0.66276	0.66640	0.67003	0.67364	0.67724	0.68082	0.68439	0.68793	0.4
0.5	0.69146	0.69497	0.69847	0.70194	0.70540	0.70884	0.71226	0.71566	0.71904	0.72240	0.5
0.6	0.72575	0.72907	0.73237	0.73565	0.73891	0.74215	0.74537	0.74857	0.75175	0.75490	0.6
0.7	0.75804	0.76115	0.76424	0.76730	0.77035	0.77337	0.77637	0.77935	0.78230	0.78524	0.7
0.8	0.78814	0.79103	0.79389	0.79673	0.79955	0.80234	0.80511	0.80785	0.81057	0.81327	0.8
0.9	0.81594	0.81859	0.82121	0.82381	0.82639	0.82894	0.83147	0.83398	0.83646	0.83891	0.9
1.0	0.84134	0.84375	0.84614	0.84849	0.85083	0.85314	0.85543	0.85769	0.85993	0.86214	1.0
1.1	0.86433	0.86650	0.86864	0.87076	0.87286	0.87493	0.87698	0.87900	0.88100	0.88298	1.1
1.2	0.88493	0.88686	0.88877	0.89065	0.89251	0.89435	0.89617	0.89796	0.89973	0.90147	1.2
1.3	0.90320	0.90490	0.90658	0.90824	0.90988	0.91149	0.91309	0.91466	0.91621	0.91774	1.3
1.4	0.91924	0.92073	0.92220	0.92364	0.92507	0.92647	0.92785	0.92922	0.93056	0.93189	1.4
1.5	0.93319	0.93448	0.93574	0.93699	0.93822	0.93943	0.94062	0.94179	0.94295	0.94408	1.5
1.6	0.94520	0.94630	0.94738	0.94845	0.94950	0.95053	0.95154	0.95254	0.95352	0.95449	1.6
1.7	0.95543	0.95637	0.95728	0.95818	0.95907	0.95994	0.96080	0.96164	0.96246	0.96327	1.7
1.8	0.96407	0.96485	0.96562	0.96638	0.96712	0.96784	0.96856	0.96926	0.96995	0.97062	1.8
1.9	0.97128	0.97193	0.97257	0.97320	0.97381	0.97441	0.97500	0.97558	0.76615	0.97670	1.9
2.0	0.97725	0.97778	0.97831	0.97882	0.97932	0.97982	0.98030	0.98077	0.98124	0.98169	2.0
2.1	0.98214	0.98257	0.98300	0.98341	0.98382	0.98422	0.98461	0.98500	0.98537	0.98574	2.1
2.2	0.98610	0.98645	0.98679	0.98679	0.98713	0.98745	0.98778	0.98809	0.98840	0.98899	2.2
2.3	0.98928	0.98956	0.89883	0.99010	0.99036	0.99061	0.99086	0.99111	0.99134	0.99158	2.3
2.4	0.99180	0.99202	0.99224	0.99245	0.99266	0.99286	0.99305	0.99324	0.99343	0.99361	2.4
2.5	0.99379	0.99396	0.99413	0.99430	0.99446	0.99461	0.99477	0.99442	0.99506	0.99520	2.5
2.6	0.99534	0.99547	0.99560	0.99573	0.99585	0.99598	0.99609	0.99621	0.99632	0.99643	2.6
2.7	0.99653	0.99664	0.99674	0.99693	0.99693	0.99702	0.99711	0.99720	0.99728	0.99736	2.7
2.8	0.99744	0.99752	0.99760	0.99767	0.99774	0.99781	0.99788	0.99795	0.99801	0.99807	2.8
2.9	0.99813	0.99819	0.99825	0.99831	0.99836	0.99841	0.99846	0.99851	0.99856	0.99861	2.9
3.0	0.99865	0.99689	0.99874	0.99878	0.99882	0.99886	0.99889	0.99893	0.99896	0.99900	3.0
3.1	0.99903	0.99906	0.99910	0.99913	0.99916	0.99918	0.99921	0.99924	0.99926	0.99929	3.1
3.2	0.99931	0.99934	0.99936	0.99938	0.99940	0.99942	0.99944	0.99946	0.99948	0.99950	3.2
3.3	0.99952	0.99953	0.99955	0.99957	0.99958	0.99960	0.99961	0.99962	0.99964	0.99965	3.3
3.4	0.99966	0.99968	0.99969	0.99970	0.99971	0.99972	0.99973	0.99974	0.99975	0.99976	3.4
3.5	0.99977	0.99978	0.99978	0.99979	0.99980	0.99981	0.99981	0.99982	0.99983	0.99983	3.5
3.6	0.99984	0.99985	0.99985	0.99986	0.99986	0.99987	0.99987	0.99988	0.99988	0.99989	3.6
3.7	0.99989	0.99990	0.99990	0.99990	0.99991	0.99991	0.99992	0.99992	0.99992	0.99992	3.7
3.8	0.99993	0.99993	0.99993	0.99994	0.99994	0.99994	0.99994	0.99995	0.99995	0.99995	3.8
3.9	0.99995	0.99995	0.99996	0.99996	0.99996	0.99996	0.99996	0.99996	0.99997	0.99997	3.9

Table 4 Percentage points of the normal distribution

The table gives the values of z satisfying $P(Z \leqslant z) = p$,
where Z is the standardised normal random variable, $N(0, 1)$.

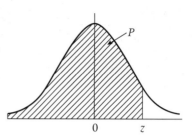

p	0.00	0.01	0.02	0.03	0.04	0.05	0.06	0.07	0.08	0.09	p
0.5	0.0000	0.0251	0.0502	0.0753	0.1004	0.1257	0.1510	0.1764	0.2019	0.2275	0.5
0.6	0.2533	0.2793	0.3055	0.3319	0.3585	0.3853	0.4125	0.4399	0.4677	0.4958	0.6
0.7	0.5244	0.5534	0.5828	0.6128	0.6433	0.6745	0.7063	0.7388	0.7722	0.8064	0.7
0.8	0.8416	0.8779	0.9154	0.9542	0.9945	1.0364	1.0803	1.1264	1.1750	1.2265	0.8
0.9	1.2816	1.3408	1.4051	1.4758	1.5548	1.6449	1.7507	1.8808	2.0537	2.3263	0.9

p	0.000	0.001	0.002	0.003	0.004	0.005	0.006	0.007	0.008	0.009	p
0.95	1.6449	1.6546	1.6646	1.6747	1.6849	1.6954	1.7060	1.7169	1.7279	1.7392	0.95
0.96	1.7507	1.7624	1.7744	1.7866	1.7991	1.8119	1.8250	1.8384	1.8522	1.8663	0.96
0.97	1.8808	1.8957	1.9110	1.9268	1.9431	1.9600	1.9774	1.9954	2.0141	2.0335	0.97
0.98	2.0537	2.0749	2.0969	2.1201	2.1444	2.1701	2.1973	2.2262	2.2571	2.2904	0.98
0.99	2.3263	2.3656	2.4089	2.4573	2.5121	2.5758	2.6521	2.7478	2.8782	3.0902	0.99

Answers

1 Introduction to statistics

EXERCISE 1A

1 (a) Qualitative;
  (b) Discrete quantitative;
  (c) Continuous quantitative (but age in years is discrete);
  (d) Continuous quantitative;
  (e) Qualitative;
  (f) Discrete quantitative;
  (g) Discrete quantitative;
  (h) Continuous quantitative;
  (i) Discrete quantitative;
  (j) Qualitative.

EXERCISE 1B

1 In this question more than one answer is possible. You may find answers in addition to those given below.

  (a) The populations mentioned in the passage will relate to all first division matches in the season and are either the results, the total number of goals scored or the amounts of time played before a goal is scored;

  (b) The total numbers of goals scored in each match played on the first Saturday of the season;

  (c) Mean number of goals per match for the season;

  (d) Mean number of goals per match on the first Saturday of the season;

  (e) The result of matches (home, away or draw);

  (f) Number of goals scored in each match (the mean number of goals scored in each match is also discrete. For example if 100 matches were played the only possible outcomes are 0.00, 0.01, 0.02 ..., however, the steps will be so small that this could also be treated as a continuous variable);

  (g) The amount of time played before a goal is scored;

  (h) The data the journalist collected in the season;

  (i) The mean number of goals per game in the previous season.

EXERCISE IC

1 (a) Place of birth, sex;

(b) Height, weight;

(c) Number of pupils weighed (age in years and months is also discrete although exact age is continuous);

(d) The data collected at the medical examination;

(e) The data collected by the class;

(f) The weights of all second year pupils;

(g) The weights of those second year pupils who were weighed;

(h) Mean weight of a sample of pupils.

2 Numerical measures

EXERCISE 2A

1 Mode 6, median 9, mean 10.

2 Mean 11.7, mode 9, median 10.

3 96 kg.

4 Mode 1, median 2, mean 2.02.

5 (a) 4; (b) 4; (c) 3.85.

6 (a) 10; (b) 10; (c) 8.42.

The highest mark is 10 (although also by far the most common). It would be more realistic if your measure of 'average' reflected the fact that no marks are greater than 10 but a substantial number of marks are less than 10. Mean is preferred.

7 {3, 3, 4, 5, 10}, {3, 3, 4, 6, 9}, {3, 3, 4, 7, 8}.

EXERCISE 2B

1 (a) 93.75 g; (b) 93.70 g.

2 (a) 62–63 s; (b) 62.5 s; (c) 62.2 s.

3 (a) 100–110; (b) 106.3 cm; (c) 105.8 cm not within tolerance.

4 (a) £176; (b) £183.

5 65.7 s.

EXERCISE 2C

1 9, 17.

2 33.5.

3 Median 19, lower quartile 16, upper quartile 21.

4 Lower quartile 9, upper quartile 16, median 13.

5 Lower quartile 9, upper quartile 15.5, median 13.

6 **(a)** 3.63 min;   **(b)** 2.21 min, 5.66 min;   **(c)** 3.44 min.

7 **(a)** £313;   **(b)** £203.

8 **(a)** $6 \leqslant x < 7$;   **(b)** £6.16;   **(c)** £1.56.

9 Mode 6, median 7, interquartile range 2.

10 {0, 1, 1, 1, 1, 1, 9} {1, 1, 1, 1, 1, 1, 8}.

EXERCISE 2D

1 3.63.

2 **A** mean 5, sd 1.58.   **B** mean 5, sd 3.16.

3 Mean 80, sd 4.63.

4 3.63 same as question 1 since variability of both samples is the same.

5 8.66 kg.

6 Mean 60.2, sd 18.8.

7 Mean 54.6 s, sd 8.33 s.

EXERCISE 2E

1 Mean 3.51, sd 1.73.

2 Mean 1.70, sd 1.61.

3 Mean 1.32, sd 1.83.

4 Mean 78.25, sd 7.66.

5 Mean 13.25, sd 5.18.

EXERCISE 2F

1 Mean 23, sd 4.

2 Mean 3.3 cm, sd 0.4 cm.

3 **(a)** Mean £51, sd £18;
   (b) Mean £31.50, sd £9.
   (c) Mean £30, sd £12.

4 Median 0.235, interquartile range 0.063.

5 **(a)** Mean 230 g, interquartile range 63 g;
   (b) Mean 223.8 g, interquartile range 60 g.

6 (a) Mean £12 400, sd £1000;

(b) Mean £12 535, sd £1090.

7 Mode 4, range 4.

EXERCISE 2G

1 A's batteries are to be preferred as they last longer and are less variable than B's.

2 (a) Line 1: mean 221, sd 104.2.
Line 2: mean 206, sd 23.9;

(b) Production line 1 has been going slightly longer on average between stoppages than production line 2. The intervals are much more variable for line 1.

3 Line 1: median 205, interquartile range 176.
Line 2: median 209, interquartile range 27.
Average as measured by median longer for line 2 but, as for the mean, the difference is small. Intervals much more variable for line 1 as in **2(b)**.

4 (a) A's ropes are strongest on average but are much more variable than B's or C's. C's ropes are much weaker than A's or B's on average. They are the least variable.

(b) High mean and low standard deviation is desirable. B is the best option.

5 (a) Moira: mean £3182, sd £1843.
Everton: mean £3522, sd £637.
Syra: mean £2592, sd £280;

(b) Syra has lowest average sales with low variability. Moira has higher average sales but is very erratic. Everton is most satisfactory with highest average sales and less variability than Moira.

6 Moira: median £2750, interquartile range £2810.
Everton: median £3435, interquartile range £1000.
Syra: median £2595, interquartile range £460.
Comparison as for **5(b)**.

7 (a) Median 361, interquartile range 507;

(b) Time to first call out for Champion is longer on average but more variable;

(c) (i) Median 526, interquartile range 893;

(ii) Average similar but now Ace is more variable;

(d) Only have to wait for 75% of machines to breakdown to calculate median and IQR. Need data for all machines to calculate mean and sd (also median and IQR less influenced by a few, very large values.)

3 Probability

Answers given as fractions or decimals are acceptable.

EXERCISE 3A

1 (a) $\dfrac{1}{20}$ or 0.05; (b) $\dfrac{5}{20}$ or $\dfrac{1}{4}$ or 0.25;

   (c) $\dfrac{6}{20}$ or $\dfrac{3}{10}$ or 0.3; (d) $\dfrac{2}{20}$ or $\dfrac{1}{10}$ or 0.1.

2 (a) $\dfrac{1}{7}$; (b) $\dfrac{2}{7}$; (c) $\dfrac{4}{7}$.

3 (a) $\dfrac{1}{7}$; (b) $\dfrac{4}{7}$; (c) $\dfrac{5}{7}$.

4 (a) $\dfrac{7}{15}$; (b) $\dfrac{1}{3}$; (c) $\dfrac{4}{5}$; (d) 0; (e) $\dfrac{1}{3}$.

5 (a) $\dfrac{1}{11}$; (b) $\dfrac{3}{11}$; (c) $\dfrac{9}{11}$; (d) $\dfrac{3}{11}$.

EXERCISE 3B

1 0.4.   2 0.78.

3 (a) 0.6; (b) 0.5; (c) 0.9.

4 (a) 0.7; (b) 0.4; (c) 0.8; (d) 0.7; (e) 0.2.

5 (a) $\dfrac{27}{35}$; (b) $\dfrac{32}{35}$; (c) $\dfrac{3}{35}$; (d) $\dfrac{23}{35}$.

6 (a) (i) *A* and *B*, (ii) *A* and *C* or *B* and *C*;

   (b) The event that the baby will not have blue eyes.

7 (a) *A*; (b) Yes; (c) *B* and *C*.

EXERCISE 3C

1 0.42.   2 0.022.   3 $\dfrac{1}{4}$ or 0.25.

4 (a) 0.01; (b) 0.81.

5 $\dfrac{1}{8}$ or 0.125.

EXERCISE 3D

Answers to three significant figures.

1 (a) 0.165; (b) 0.48; (c) 0.615.

2 (a) (i) 0.846, (ii) 0.147, (iii) 0.154;

   (b) (i) 0.779, (ii) 0.203, (iii) 0.0182.

3 (a) (i) 0.0625; (ii) 0.375;

   (b) (i) 0.422, (ii) 0.141, (iii) 0.156, (iv) 0.844.

   (c) 0.316.

4 (a) (i) 0.64, **(ii)** 0.24, **(iii)** 0.665,
(iv) 0.9025, **(v)** 0.335;

(b) (i) 0.512, **(ii)** 0.288, **(iii)** 0.008,
(iv) 0.5155, **(v)** 0.036.

EXERCISE 3E

1 (a) $\dfrac{5}{8}$; **(b)** $\dfrac{2}{3}$; **(c)** $\dfrac{3}{8}$; **(d)** $\dfrac{47}{120}$; **(e)** $\dfrac{9}{10}$;

(f) $\dfrac{47}{80}$; **(g)** $\dfrac{28}{75}$; **(h)** $\dfrac{7}{10}$; **(i)** $\dfrac{33}{80}$; **(j)** $\dfrac{7}{30}$.

2 (a) $\dfrac{23}{30}$; **(b)** $\dfrac{121}{150}$; **(c)** $\dfrac{25}{29}$; **(d)** $\dfrac{119}{150}$; **(e)** $\dfrac{1}{6}$;

(f) $\dfrac{3}{5}$; **(g)** $\dfrac{2}{5}$; **(h)** $\dfrac{90}{121}$; **(i)** $\dfrac{31}{35}$.

3 (a) $\dfrac{4}{9}$; **(b)** $\dfrac{2}{3}$; **(c)** $\dfrac{5}{9}$; **(d)** $\dfrac{2}{15}$; **(e)** $\dfrac{4}{5}$;

(f) $\dfrac{13}{23}$; **(g)** $\dfrac{1}{5}$; **(h)** $\dfrac{7}{10}$; **(i)** 0.

(j) A and C not independent.

EXERCISE 3F

Answers to three significant figures.

1 (a) 0.195; **(b)** 0.499.

2 (a) 0.357; **(b)** 0.536.

3 (a) (i) 0.0152, **(ii)** 0.182, **(iii)** 0.227;
(b) (i) 0.0909, **(ii)** 0.136, **(iii)** 0.409, **(iv)** 0.218.

4 (a) 0.0480; **(b)** 0.506; **(c)** 0.305.

5 (a) 0.196; **(b)** 0.0240; **(c)** 0.0840; **(d)** 0.0960;
(e) 0.240; **(f)** 0.228; **(g)** 0.192.

6 (a) 0.0461; **(b)** 0.233; **(c)** 0.0121; **(d)** 0.279;
(e) 0.0101; **(f)** 0.266; **(g)** 0.152.

MIXED EXERCISE

1 (a) $\dfrac{1}{343}$; **(b)** $\dfrac{1}{49}$; **(c)** $\dfrac{30}{49}$; **(d)** $\dfrac{8}{343}$; **(e)** 6.

2 (a) (i) 0.576, **(ii)** 0.932; **(b)** 0.912.

3 (a) 0.0429; **(b)** 0.142; **(c)** 0.1215; **(d)** 0.189;
(e) 0.334.

4 (a) (i) 0.36, **(ii)** 0.09, **(iii)** 0.89, **(iv)** 0.36;
(b) $R' \cap T'$ [or $(R \cup T)'$].

5 (a) (i) 0.343, **(ii)** 0.441,

  (b) (i) 0.063, **(ii)** 0.09; **(c)** 0.141.

6 (a) (i) 0.12, **(ii)** 0.0455, **(iii)** 0.318, **(iv)** 0.0133,

  (v) 0.893;

  (b) 0.0827.

7 (a) (i) 0.462, **(ii)** 0.223, **(iii)** 0.808, **(iv)** 0.517;

  (b) (i) 0.1575, **(ii)** 0.153;

  (c) 0.224.

4 Binomial distribution

Answers given as fractions or decimals are acceptable.

EXERCISE 4A

1

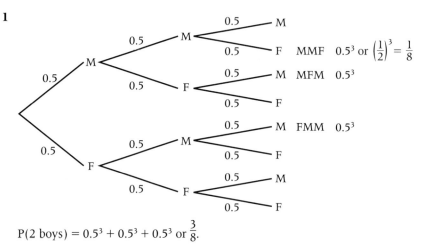

$P(2 \text{ boys}) = 0.5^3 + 0.5^3 + 0.5^3 \text{ or } \dfrac{3}{8}.$

2

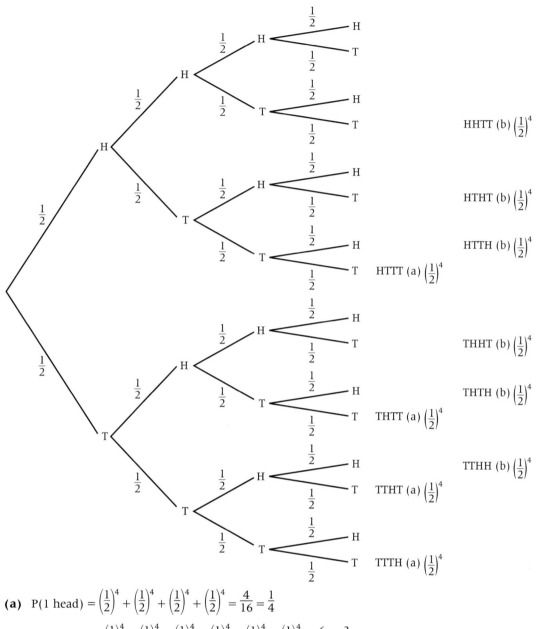

HHTT (b) $\left(\frac{1}{2}\right)^4$

HTHT (b) $\left(\frac{1}{2}\right)^4$

HTTH (b) $\left(\frac{1}{2}\right)^4$

HTTT (a) $\left(\frac{1}{2}\right)^4$

THHT (b) $\left(\frac{1}{2}\right)^4$

THTH (b) $\left(\frac{1}{2}\right)^4$

THTT (a) $\left(\frac{1}{2}\right)^4$

TTHH (b) $\left(\frac{1}{2}\right)^4$

TTHT (a) $\left(\frac{1}{2}\right)^4$

TTTH (a) $\left(\frac{1}{2}\right)^4$

(a) $\text{P(1 head)} = \left(\frac{1}{2}\right)^4 + \left(\frac{1}{2}\right)^4 + \left(\frac{1}{2}\right)^4 + \left(\frac{1}{2}\right)^4 = \frac{4}{16} = \frac{1}{4}$

(b) $\text{P(2 heads)} = \left(\frac{1}{2}\right)^4 + \left(\frac{1}{2}\right)^4 + \left(\frac{1}{2}\right)^4 + \left(\frac{1}{2}\right)^4 + \left(\frac{1}{2}\right)^4 + \left(\frac{1}{2}\right)^4 = \frac{6}{16} = \frac{3}{8}.$

3 $0.4 \times 0.4 \times 0.4 = 0.064$
not   not   not
light light light

4

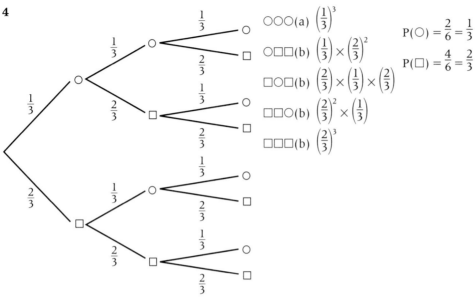

$P(\bigcirc) = \frac{2}{6} = \frac{1}{3}$

$P(\square) = \frac{4}{6} = \frac{2}{3}$

(a) $P(\text{no squares}) = \frac{1}{3} \times \frac{1}{3} \times \frac{1}{3} = \frac{1}{27}$;

(b) $P(\text{2 or 3 squares}) = \frac{1}{3} \times \left(\frac{2}{3}\right)^2 + \frac{1}{3} \times \left(\frac{2}{3}\right)^2 + \frac{1}{3} \times \left(\frac{2}{3}\right)^2 + \left(\frac{2}{3}\right)^3$

$= \frac{4}{27} + \frac{4}{27} + \frac{4}{27} + \frac{8}{27}$

$= \frac{20}{27}$

(At least two squares means two or three.)

EXERCISE 4B

1 $n = 4$, $p = \frac{1}{4}$, $P(X = 2) = \binom{4}{2}\left(\frac{1}{4}\right)^2\left(\frac{3}{4}\right)^2 = 6 \times \frac{1}{16} \times \frac{9}{16} = \frac{27}{128}$.

2 $n = 5$, $p = \frac{3}{5}$, $P(X = 4) = \binom{5}{4}\left(\frac{3}{5}\right)^4\left(\frac{2}{5}\right)^1 = 5 \times \frac{81}{625} \times \frac{2}{5} = \frac{162}{625}$.

3 $n = 10$, $p = 0.4$, $P(X = 3) = \binom{10}{3}\left(0.4\right)^3\left(0.6\right)^7 = 120 \times 0.064 \times 0.6^7$

$= 0.215$.

4 $n = 10$, $p = 0.2$, $P(X = 3) = \binom{10}{3}\left(0.2\right)^3\left(0.8\right)^7 = 120 \times 0.008 \times 0.8^7$

$= 0.201$.

5 $n = 10$, $p = 0.3$, $P(X = 4) = \binom{10}{4}\left(0.3\right)^4\left(0.7\right)^6 = 210 \times 0.3^4 \times 0.7^6$

$= 0.200$.

6 $n = 20$, $p = 0.15$, $P(X = 2) = \binom{20}{2}(0.15)^2(0.85)^{18}$

$= 190 \times 0.15^2 \times 0.85^{18} = 0.229$.

7 $n = 30$, $p = 0.2$, $P(X = 8) = \binom{30}{8}(0.2)^8(0.8)^{22} = 5852925 \times 0.2^8 \times 0.8^{22}$
$$= 0.111.$$

8 $n = 25$, $p = 0.1$, $P(X = 2) = \binom{25}{2}(0.1)^2(0.9)^{23} = 300 \times 0.01 \times 0.9^{23}$
$$= 0.266.$$

EXERCISE 4C

Answers from tables have been given to four decimal places. However three significant figures is sufficient.

1 $n = 20$, $p = 0.2$, $X \sim B(20, 0.2)$

   (a) $P(X \leq 3) = 0.4114$;

   (b) $P(X < 3) = P(X \leq 2) = 0.2061$;

   (c) $P(X > 1) = 1 - P(X \leq 1) = 1 - 0.0692 = 0.9308$.

0, 1, 2, 3, |4 ...

0, 1, 2,| 3, 4 ...

0, 1,| 2, 3, 4 ...

X ≤ 1   X > 1

2 $n = 8$, $p = \frac{2}{5}$, $X \sim B(8, 0.4)$

   (a) $P(X < 3) = P(X \leq 2) = 0.3154$;

   (b) $P(X \geq 2) = 1 - P(X \leq 1) = 1 - 0.1064 = 0.8936$;

   (c) $P(X = 0) = 0.0168$.

0, 1, 2,| 3, 4 ...

0, 1,| 2, 3, 4 ...

X ≤ 1   X ≥ 2

3 $n = 25$, $p = \frac{1}{5} = 0.2$, $X \sim B(25, 0.2)$

   (a) $P(X > 5) = 1 - P(X \leq 5) = 1 - 0.6167 = 0.3833$;

   (b) $P(X \geq 6) = 1 - P(X \leq 5) = 0.3833$
      at least 6 = 6 or more = more than 5 see **(a)**;

   (c) $P(X < 4) = P(X \leq 3) = 0.2340$.

0, 1, 2, 3, 4, 5,|6, 7 ...

X ≤ 5   X > 5

0, 1, 2, 3,|4, 5, 6, 7 ...

4 $n = 25$, $p = 0.3$, $X \sim B(25, 0.3)$

   (a) $P(X < 5) = P(X \leq 4) = 0.0905$;

   (b) $P(X \leq 8) = 0.6769$
      no more than 8 = 8 or less;

   (c) $P(X > 3) = 1 - P(X \leq 3) = 1 - 0.0332 = 0.9668$.

0, 1, 2, 3, 4,|5, 6, 7, 8

0, 1, 2, 3, 4, 5, 6, 7, 8,|9

0, 1, 2, 3,|4, 5, 6, 7, 8, 9 ...

X ≤ 3   X > 3

5 $n = 20$, $p = \frac{3}{100} = 0.03$, $X \sim B(20, 0.03)$

   (a) $P(X = 0) = 0.5438$;

   (b) $P(X \geq 2) = 1 - P(X \leq 1) = 1 - 0.8802 = 0.1198$;

   (c) $P(X < 3) = P(X \leq 2) = 0.9790$;

   (d) $P(X = 1) = P(X \leq 1) - P(X = 0) = 0.8802 - 0.5438 = 0.3364$

      or $\binom{20}{1}(0.03)^1(0.97)^{19} = 0.336$ (three significant figures).

0, 1,| 2, 3, 4

0, 1, 2,| 3, 4

0, 1,| 2, 3, 4

6 $n = 25$, $p = 0.15$, $X \sim B(25, 0.15)$

   (a) $P(X \geq 4) = 1 - P(X \leq 3) = 1 - 0.4711 = 0.5289$;

   (b) $P(X \leq 5) = 0.8385$
      no more than 5 = 5 or less;

   (c) $P(5 \leq X \leq 10) = P(X \leq 10) - P(X \leq 4) = 0.9995 - 0.6821 = 0.3174$.

0, 1, 2, 3,|4, 5, 6, 7, 8, 9,10 ...

X ≤ 3 X ≥ 4

0, 1, 2, 3, 4,|5, 6, 7, 8, 9,10,|11 ...

X ≤ 4   5 ≤ X ≤ 10

7 $n = 12, p = 0.2, X \sim B(12, 0.2)$

$\overline{0, 1, 2,}3,|4, 5$

   (a) $P(X \leqslant 3) = 0.7946$;

   (b) $P(X = 3) = 0.2363 = P(X \leqslant 3) - P(X \leqslant 2)$, (direct calculation gives 0.2362);

   (c) $P(X \leqslant 1) = 0.2749$
     no more than $1 = 1$ or fewer;

   (d) $P(\geqslant 10$ agree$) = P(10, 11, 12$ agree$) = P(0, 1, 2$ refuse$)$
     $P(X \leqslant 2) = 0.5583$;

8 $n = 40, p = 0.15, X \sim B(40, 0.15)$

$\overline{0, 1, 2, 3, 4, 5, 6,}7,|8 \dots$

   (a) $P(X \leqslant 5) = 0.4325$;

   (b) $P(X = 7) = P(X \leqslant 7) - P(X \leqslant 6) = 0.7359 - 0.6067 = 0.1492$
     (direct calculations gives 0.1493);

$\overline{0, 1, 2, 3,}|4, 5, 6, 7, 8, 9, 10,|11$

$X \leqslant 3 \qquad 4 \leqslant X \leqslant 10$

   (c) $P(4 \leqslant X \leqslant 10) = P(X \leqslant 10) - P(X \leqslant 3) = 0.9701 - 0.1302 = 0.8399$;

   (d) $P(36, 37, 38, 39, 40$ agree$) = P(0, 1, 2, 3, 4$ decline$)$
     $P(X \leqslant 4) = 0.2633$.

9 **(a)** 0.0644;   **(b)** 0.9956;   **(c)** 0.0190.

10 **(a)** **(i)** 0.512,   **(ii)** 0.384;

     (b) **(i)** 0.491,   **(ii)** 0.421.

EXERCISE 4D

Answers from tables have been given to 4 d.p. However, 3 significant figures is sufficient.

1 **(a)** $W \sim B(20, 0.08)$

     (i) $P(W = 0) = 0.1887$,

     (ii) $P(W \geqslant 2) = 1 - P(W \leqslant 1) = 1 - 0.5169 = 0.4831$,

     (iii) $P(W = 2) = P(W \leqslant 2) - P(W \leqslant 1) = 0.7879 - 0.5169 = 0.2710$;

   (b) $R \sim B(22, 0.08)$

$$P(R \geqslant 2) = 1 - P(R = 1) - P(R = 0)$$
$$= 1 - \binom{22}{1}(0.08)(0.92)^{21} - (0.92)^{22}$$
$$= 1 - 0.3055 - 0.1597 = 0.5348.$$

2 **(a)** $n = 40, p = 0.15, X \sim B(40, 0.15)$

     (i) $P(X \leqslant 5) = 0.4325$,

     (ii) $P(X = 7) = P(X \leqslant 7) - P(X \leqslant 6) = 0.7559 - 0.6067 = 0.1492$,

     (iii) $P(X > 4) = 1 - P(X \leqslant 4) = 1 - 0.2633 = 0.7367$;

$0, 1, 2, 3, 4,|\overline{5, 6, \dots}$

$X \leqslant 4 \qquad X > 4$

   (b) $n = 25, p = 0.2, X \sim B(25, 0.2)$

     (i) $P(X \geqslant 3) = 1 - P(X \leqslant 2) = 1 - 0.0982 = 0.9018$,

     (ii) $P(X \leqslant 5) = 0.6167$;

$0, 1, 2,|\overline{3, 4 \dots}$

$X \leqslant 2 \quad X \geqslant 3$

   (c) X is not binomial, since the total number of trials is not fixed.

3 $n = 25, p = \dfrac{1}{5} = 0.2, X \sim B(25, 0.2)$

   (a) **(i)** $P($marks $\leqslant 8) = P(X \leqslant 8) = 0.9532$

     (ii) $P($marks $> 12) = P(X > 12) = 1 - P(X \leqslant 12)$
                      $= 1 - 0.9996 = 0.0004$,

$0, \dots, 10, 11, 12,|\overline{13}$

$X \leqslant 12 \quad X \geqslant 13$

     (iii) $P(10$ marks$) = P(X = 10) = P(X \leqslant 10) - P(X \leqslant 9)$
                       $= 0.9944 - 0.9827 = 0.0117$ (direct calculation gives 0.0118);

(b) Mean mark = $np = 25 \times 0.2 = 5$

   (i) 0,

   (ii) 75

4 $X \sim B(20, 0.4)$

   (a) $P(X \geq 10) = 1 - P(X \leq 9) = 1 - 0.7553 = 0.2447;$

   (b) $P(\text{yes}) = 0.6$ too big for tables

   $P(\text{no}) = 0.4$ use $p = 0.4$

   $P(\geq 10 \text{ say 'yes'})$

   $= P(\leq 10 \text{ say 'no'}) \; X \sim B(20, 0.4)$ for 'no'

   $= 0.8725;$

10, 11, 12, ..., 19, 20, Yes	
10, 9, 8, ..., 1, 0, No	

   (c) Union meeting is likely to influence opinions.
   Probability of voting 'Yes' may be different for those attending meeting.
   Trials are not independent as a show of hands is used. Drivers may be influenced by how friends vote.

5 (a) (i) 0.3823,

   (ii) 0.3669;

   (b) 4, 1.55;

   (c) (i) 3.92, 3.17 (or 3.05),

   (ii) mean similar, sd much greater
   → binomial not good model
   → Siballi's beliefs not plausible.

6 (a) (i) 0.0433,

   (ii) 0.4395;

   (b) No, n not constant;

   (c) p may change (decrease) as Dwight gains experience

5 The normal distribution

Most answers are rounded to 3 s.f. and can vary slightly dependent on whether tables or a calculator are used.

EXERCISE 5A

1 (a) 0.891;   **(b)** 0.834;   **(c)** 0.968;

   (d) 0.663;   **(e)** 0.536;   **(f)** 0.942;

   (g) 0.974;   **(h)** 0.726;   **(i)** 0.991;

   (j) 0.853.

EXERCISE 5B

1 (a) 0.0869;   **(b)** 0.281;   **(c)** 0.109;

   (d) 0.195;   **(e)** 0.374;   **(f)** 0.0262;

   (g) 0.008 89;   **(h)** 0.258;   **(i)** 0.464;

   (j) 0.0885.

EXERCISE 5C

1 (a) 0.0823;   (b) 0.281;   (c) 0.862;

  (d) 0.681;   (e) 0.326;   (f) 0.626;

  (g) 0.261;   (h) 0.802;   (i) 0.773;

  (j) 0.330.

EXERCISE 5D

1 (a) 0.209;   (b) 0.0948;   (c) 0.516;

  (d) 0.877;   (e) 0.112;   (f) 0.0214;

  (g) 0.888;   (h) 0.003 61;   (i) 0.0792;

  (j) 0.968.

EXERCISE 5E

1 (a) 1.4;   (b) 0.6;   (c) −0.8;

  (d) −1.6;   (e) 2.1.

2 (a) 0.652;   (b) −1.370;   (c) 1.348;   (d) −1.804.

3 (a) 1.167;   (b) −0.167;   (c) 0.667;   (d) −1.167.

EXERCISE 5F

1 (a) 0.663;   (b) 0.104;   (c) 0.134;

  (d) 0.943;   (e) 0.396;   (f) 0.256.

2 (a) 0.779;   (b) 0.663;   (c) 0.0336;

  (d) 0.0367;   (e) 0.913;   (f) 0.295.

3 (a) 0.245;   (b) 0.755;   (c) 0.832;

  (d) 0.152;   (e) 0.659;   (f) 0.603.

4 (a) 0.637;   (b) 0.0838;   (c) 0.0455;

  (d) 0.705;   (e) 0.179;   (f) 0.395.

5 (a) 0.230;   (b) 0.152;   (c) 0.858.

EXERCISE 5G

1 (a) +1.960;   (b) −1.282;   (c) −1.645;

  (d) +1.44 (using interpolation);   (e) −1.960;

  (f) +1.036;   (g) −0.842;   (h) +1.282;

  (i) +2.326.

2 (a) −1.645 and +1.645;

  (b) −2.576 and +2.576;

  (c) −3.090 and +3.090.

EXERCISE 5H

1 (a) 85.5; (b) 80.1; (c) 69.3;

   (d) 88.8; (e) 59.2–88.8; (f) 66.4–81.6;

2 (a) 421; (b) 370; (c) 355;

   (d) 321; (e) 231; (f) 277–433.

3 (a) 2.35 p.m.; (35 minutes past 2 – answer to nearest minute to ensure arriving **before** 3.00);

   (b) 2.30 p.m.;

   (c) 2.26 p.m.;

   (d) 2.46 p.m.;

   (e) 2.40 p.m. (nearest minute to arrive **after** 3.00 p.m.);

   (f) 2.47 p.m.

EXERCISE 5I

1 26.78–31.62 cm.

2 (a) (i) 0.894, (ii) 0.493;

   (b) Longest possible stay is 60 minutes. For proposed model about 60% of times will exceed 60 minutes. Model could not apply;

   (c) 99.9% of normal distribution less than $\mu + 3\sigma$, i.e. $65 + 3 \times 20 = 125$ minutes. Model could be plausible for users entering 125 minutes or more before closing time, i.e. 6.55 p.m.

   Note: this answer is very cautious, you could argue that 95% of the distribution is sufficient.

3 (a) 0.405; (b) £761; (c) 0.0269;

   (d) Cannot be exact because money is a discrete variable and also because negative takings impossible.

EXERCISE 5J

Interpolation has been used, your answers may be slightly different if you have not used interpolation.

1 (a) 0.909; (b) 0.0710; (c) 0.838.

2 (a) 45.75 – 46.65 cm;

   (b) 46.58 cm;

   (c) 128.

3 (a) (i) 0.122, (ii) 0.661, (iii) 0.488;

   (b) 15.6 – 20.4 s.

4 (a) 129;

   (b) Large sample $\rightarrow$ sample mean normally distributed, whatever the population distribution;

   (c) Might be invalid if sample size small and population not normal (or sample not random).

MIXED EXERCISE

1 **(a)** 0.291 or 29.1%;   **(b)** 403.

2 0.841.

3 **(a)** 0.159;   **(b)** 0.801;   **(c)** 60.2 g.

4 **(a)** 0.919;   **(b)** 0.274;   **(c)** 8.11 a.m.

5 **(a)** 0.022 75;

   (b) 0.821,

   (c) 0.0886 (answers may be slightly different without interpolation). Large sample in **(c)** so answer unchanged if weights not normally distributed.

6 0.0475 or 4.75%.

7 0.910.

8 **(a)** 0.401;   **(b)** 0.691;   **(c)** 0.494, unaffected.

9 **(a)** 0.212 or 21.2%;   **(b)** 22.6 g.

10 **(a)** 0.386;   **(b)** 0.0644. Money is discrete variable normal is continuous; cannot carry a negative amount of change.

11 **(a)** 0.115;   **(b)** 33.2 g.

12 **(a)** **(i)** 0.022 75,   **(ii)** 0.440,   **(iii)** 0.785;

   (b) **(i)**   Mean only 1.08 sd above zero. For normal this gives probability of about 0.14 of negative times, which are impossible,

      (ii) Large sample → mean approximately normally distributed.

13 **(a)** **(i)** 0.655,   **(ii)** 0.444;

   (b) 0.736;

   (c) 0.115;

   (d) 423 ml;

   (e) 9.65 ml.

14 **(a)** 0.685;

   (b) 40.5 m;

   (c) 40 m, 4.9 m;

   (d) Gwen. Yuk Ping has a negligible chance of throwing 48 m.

15 **(a)** **(i)** 0.0495,   **(ii)** 0.000 05; **I** and **II** aren't satisfied.

   (b) **(i)** 105.3 ml,   **(ii)** 106.5 ml, mean at least 106.5 ml;

   (c) **(i)**   103.3 ml (mean), 3.98 ml (sd),

      (ii) Yes, conditions just met, but if sd reduced, conditions can be met with smaller mean content.

6 Confidence intervals

Answers are rounded. More than 3 s.f. are given where appropriate.

EXERCISE 6A

1 (a) 58.64–60.00;   (b) 58.81–59.83;   (c) 59.10–59.54.

2 (a) (i) 65.0–82.2,   (ii) 63.3–83.8,   (iii) 60.1–87.0;

  (b) (i) 83.9–97.3,   (ii) 80.4–100.8,   (iii) 77.1–104.1;

  (c) Athletes seem to have a lower mean diastolic blood pressure than
  for the population of healthy adults (84.8 is above the 95%
  confidence interval, although it is just inside the 99% interval). On
  this evidence chess club members are consistent with the
  population of healthy adults as 84.8 lies within the confidence
  intervals.

3 63.5–115.5.

4 (a) 101.7–159.2;   (b) 29.3–174.1;

  (c) Station manager's claim is incorrect. Even making the lowest
  reasonable estimate of the mean the great majority of passengers
  will queue for more than 25 s.

5 (a) 494.69–499.63;   (b) (i) 7.5 g,   (ii) 440.3–478.9;

  (c) Confidence interval calculated in (a) suggests that the mean
  weight of pickles in a jar is above 454 g but interval calculated in
  (b) suggests that many individual jars will contain less than 454 g
  of pickles.

EXERCISE 6B

1 £93.6–£101.4.

2 (a) 812.2–883.8 g;   (b) 831.3–864.7 g.

3 (a) 72.54–75.46 cm;

  (b) No difficulty as sample is large so mean will be approximately
  normally distributed.

4 (a) (i) 2.717–2.755 cm,   (ii) 2.679–2.701 cm;

  (b) 2.608–2.772 cm;

  (c) Confidence intervals do not overlap so mean for soft centres clearly
  greater than mean for hard centres. However interval calculated in
  (b) shows that many hard-centred chocolates are bigger than the
  mean of the soft-centred chocolates. Diameter not a great deal of
  use because of large amount of overlap.

5 (a) 201.35–208.51 g;   (b) (i) 0.150,   (ii) 0.0265;

  (c) Average weight okay, too large a proportion less than 191 g and
  less than 182 g. This could be rectified by increasing the mean.
  Meeting the requirements in this way will mean that the mean
  contents are quite a lot over the nominal weight. Reducing the
  standard deviation is expensive but would allow the
  requirements to be met with a small reduction in the current
  mean contents.

EXERCISE 6C

1 (a) 266–516 days; **(b)** 0.1.

2 (a) (i) 2640–3080 hours, **(ii)** 139;

  (b) (i) Some uncertainty as sample is small,

    (ii) No problem as sample is large;

  (c) Some doubt as if standard deviation is 300, the sample range is only about 1.5 standard deviations (or estimated standard deviation only 184).

3 (a) (i) 193.1–200.1 g, **(ii)** 195.1–198.1 g;

  (b) (i) 0.05, **(ii)** 0.05;

  (c) 0.02.

4 (a) 925–1092 days;

  (b) No substantial evidence of any difference;

  (c) 167; **(d)** 75.8; **(e)** 20.

5 (a) 46.8–49.4 hours; **(b)** 2.67;
  (c) 85.8%; **(d)** 0.142.

6 (a) 311–943 ml;

  (b) (i) Distribution appears skew (3 zeros),

    (ii) amount cannot be negative;

  (c) (i) 926–994 ml,

    (ii) large sample $\rightarrow$ mean normally distributed,

    (iii) first day unlikely to be typical/night shift may differ from day shift.

7 (a) (i) 496.7–500.5 g, **(ii)** 497.4–499.8 g;

  (b) (i) (A) 0.8; **(B)** 0.15, **(ii)** 0.19.

8 (a) 36.3–60.5;

  (b) (i) 46.42–52.58, **(ii)** 17.2–81.8;

  (c) Evidence mean exceeds 25, but some individuals will score less than 25;

  (d) Interval narrower, no need for normal assumption;

  (e) Both samples random so confidence interval valid. Estimate better but not much.

7 Correlation

EXERCISE 7A

1 (a) (i) Yes, **(ii)** Yes, **(iii)** No;

  (b) (i) No – all points do not lie on the same line,

    (ii) Yes – weak, negative correlation is evident,

    (iii) No – $-1 \le r \le 1$ so $r = 1.2$ is impossible.

2 Answers to this question are approximate.

   (i) 0.9;

   (ii) −0.3;

   (iii) 0.85;

   (iv) 0.

3 **(a)** and **(b)**   **(i)**

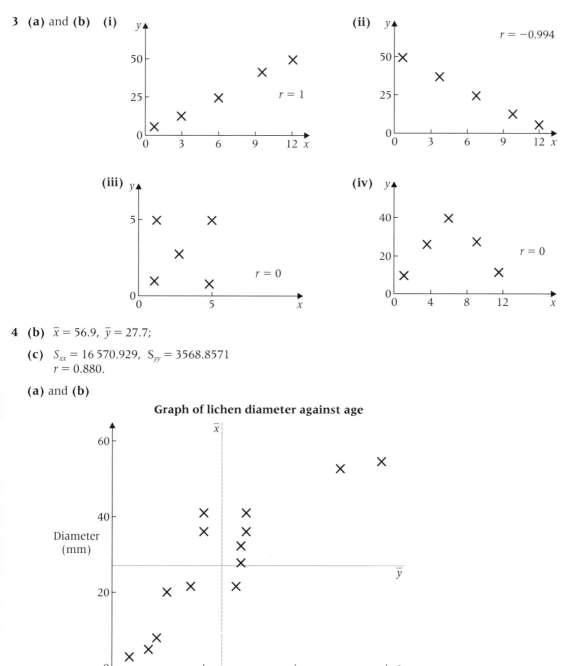

4 **(b)** $\bar{x} = 56.9$, $\bar{y} = 27.7$;

   (c) $S_{xx} = 16\,570.929$, $S_{yy} = 3568.8571$
   $r = 0.880$.

   (a) and **(b)**

5 (a)   **Graph of length against temperature**

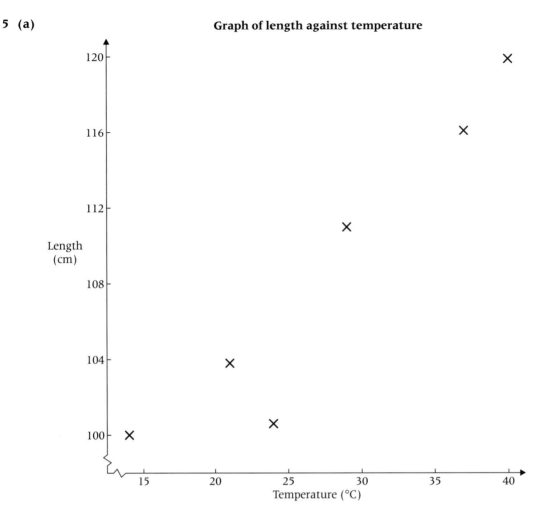

(b) $r = 0.949$.

(c) Discard (24, 100.6)) → new $r = 0.996$.
It appears there is an almost exact linear relationship once (24, 100.6) is removed. However, even when including this point the fit was very good.

6 (a)

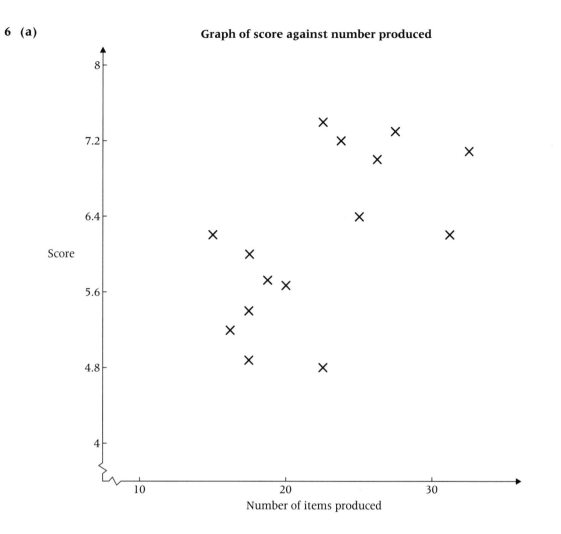

Graph of score against number produced

(b) $r = 0.610$;

(c) The data appear to show that the owner's belief is incorrect as there is weak positive correlation. However, if the data were divided into two groups, the owner's belief may be true as the experienced craftsmen (2, 4, 8, 9, 10, 13, 14 perhaps) may produce higher quality goods, but this quality may deteriorate if rushed.

7 (a)

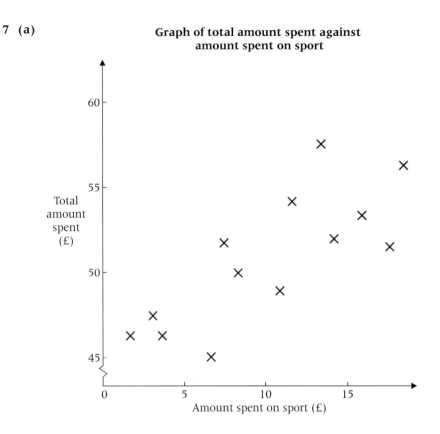

**Graph of total amount spent against
amount spent on sport**

(b) $r = 0.824$;

(c) It appears appropriate since the data seem to follow a linear
relationship. Since x is part of y, it might be better to examine
relationship of x with $y - x$.

8 (b) $r = 0.937$;

(c) Calculations seem inappropriate as a clear non-linear relationship
is seen – despite high value of r.

9 (a) $r = -0.798$;

(b) $r = -0.817$;

(c) Both values for r show fairly strong negative correlation indicating
(a) higher heart value function links to lower baldness and **(b)**
higher hours of TV links to lower heart function.

(d) Data does not provide evidence for a causal link between watching
TV and any effect on heart function. There may be a separate
factor which is linked to both number of hours watching TV and
heart function.

8 Regression

EXERCISE 8A _____

1 **(a)**

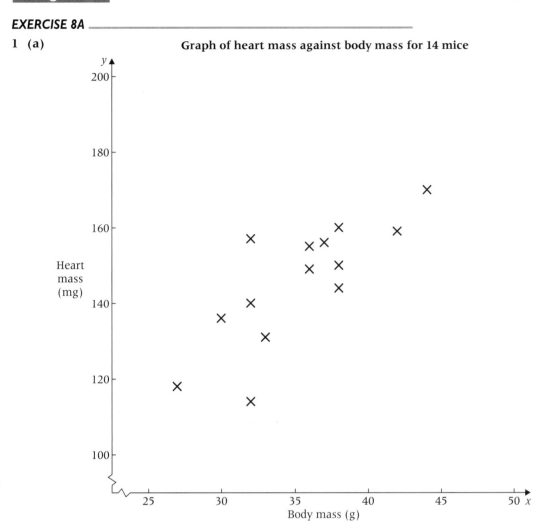

Graph of heart mass against body mass for 14 mice

(b) $y = 48.4 + 2.75x$.

2 (a)

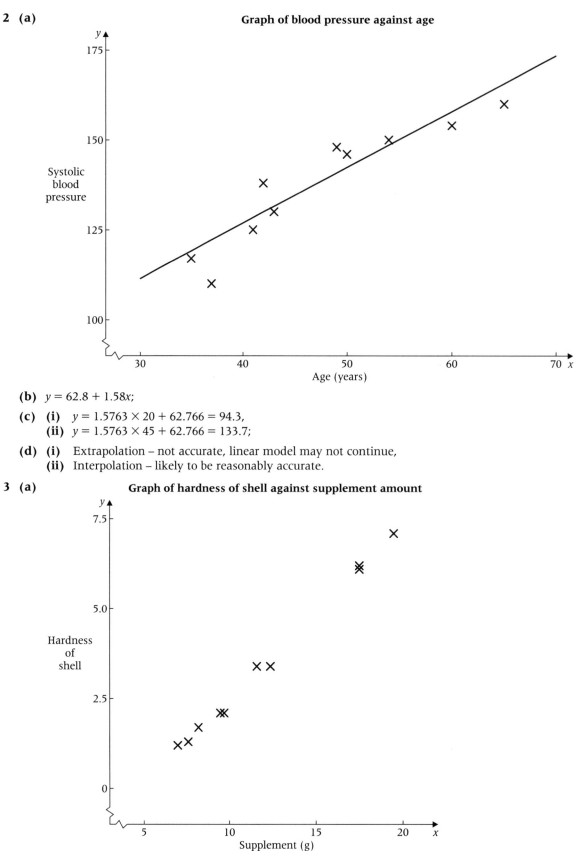

Graph of blood pressure against age

(b) $y = 62.8 + 1.58x$;

(c) (i) $y = 1.5763 \times 20 + 62.766 = 94.3$,
(ii) $y = 1.5763 \times 45 + 62.766 = 133.7$;

(d) (i) Extrapolation – not accurate, linear model may not continue,
(ii) Interpolation – likely to be reasonably accurate.

3 (a)

Graph of hardness of shell against supplement amount

(b) $y = 0.486x - 2.40$;

(c) y on scale 0–10, model cannot extend to values of x outside range 5–25 (approximately).

4 (a) and **(c)**

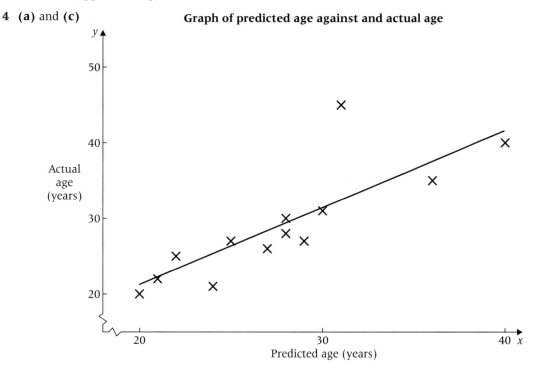

Graph of predicted age against and actual age

(b) $y = 1.03x + 0.533$;

(d) With the exception of G, predictions seem fairly accurate – the points all lie close to the line. It would be advisable to investigate person G to see if they should be excluded (been ill/in prison?).

5 **(a)** and **(c)**

Graph of price against capacity

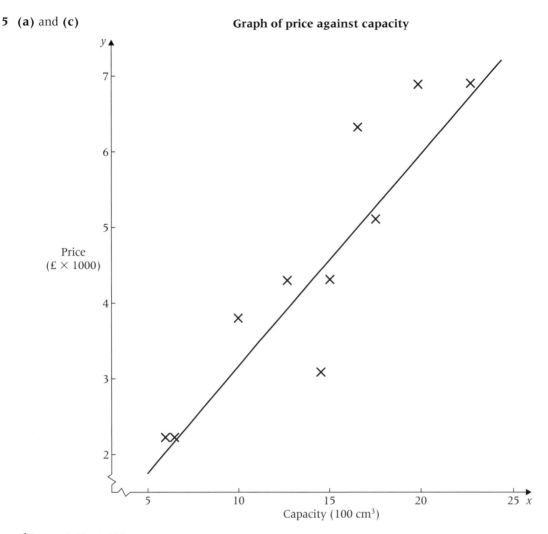

(b) $y = 3.02x + 237$;

(d) Model J is recommended (well below line – very low price).
Discourage models A, E and K (above line – high price)

6 (a)

Graph of salary against score

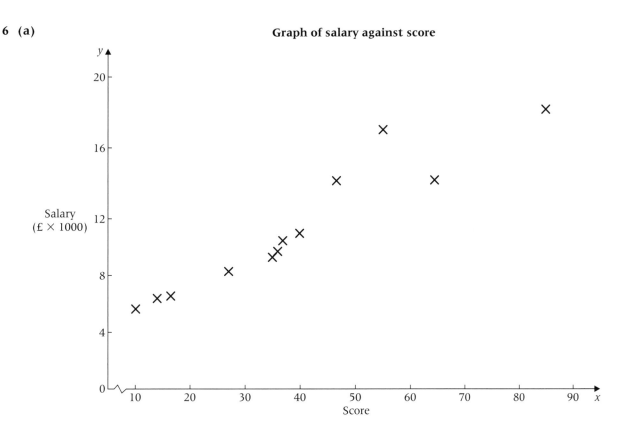

(b) $y = 192x + 3713$;

(c) Points close to straight line, apart from *B* and *C*. Method should be reasonably satisfactory.

(d) Salary $= a + bx + t$, where t is an additional payment for employees who have to work away from home.

7 (a)

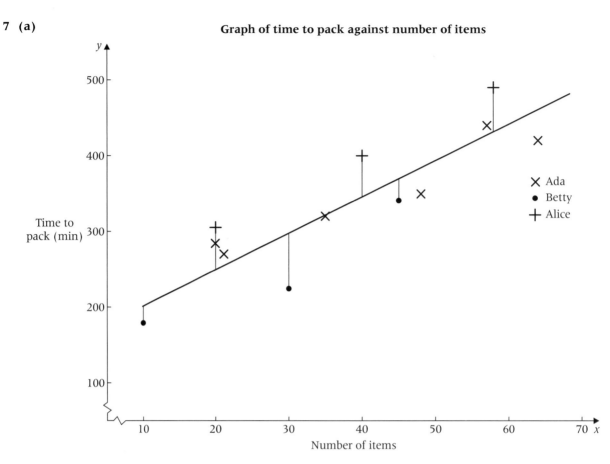

Graph of time to pack against number of items

(b) $r = 0.897$, consistent with linear relationship, more items $\rightarrow$ more time to pack;

(c) $y = 166 + 4.62x$;

(d) $y = 4.6161 \times 45 + 165.52 = 373$, should be fairly accurate but would depend on packer;

(e) Betty $\simeq -59.0, -31.7, -33.2$, average -41.3,
Alice $\simeq +47.2, +49.8, +56.7$, average $+51.2$;

(f) **(i)** Betty $373.2 - 41.3 = 332$,
   (ii) Alice $373.2 + 51.2 = 424$.

8 (a)

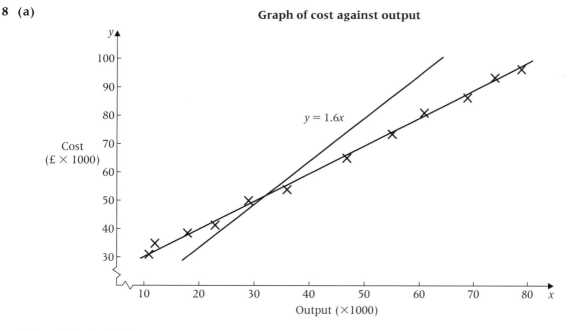

Graph of cost against output

$y = 1.6x$

Cost
(£ × 1000)

Output (×1000)

(b) $y = 0.961x + 20.7$;

(c) Approximately 32 000 output;

(d) If output above 32 000 a quarter, profit will be made.

9 (a)

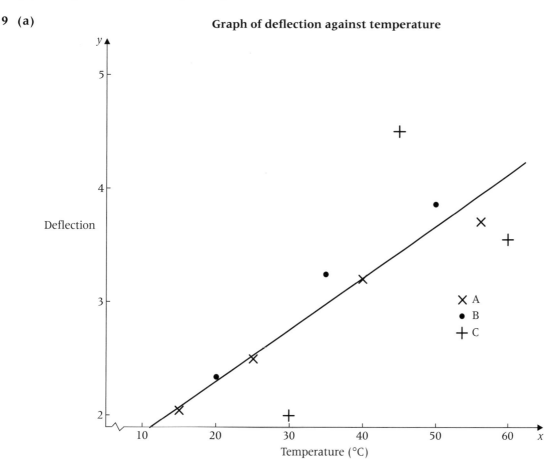

Graph of deflection against temperature

Deflection

× A
• B
+ C

Temperature (°C)

(b) $y = 0.0453x + 1.42$;

(c) Technician B seems to give higher results than A. A and B are both consistent, however, C's results are very erratic;

(d) Check which of A and B is 'accurate'. Try to find and eliminate cause of small systematic difference between A and B. Check C's measurements: C needs retraining.

10 **(a)** and **(b)**   **Graph of takings against number of part-time staff**

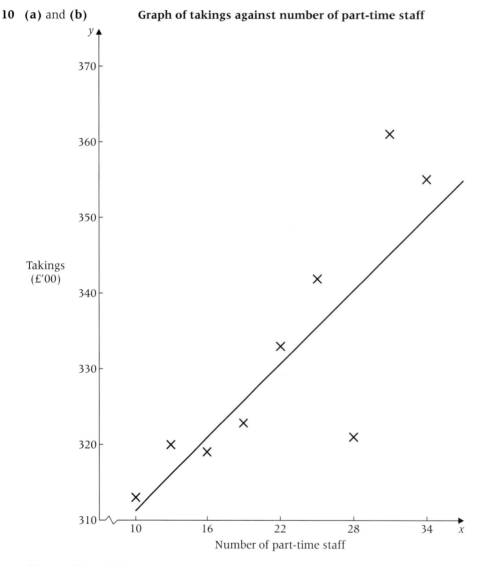

(b) $y = 294 + 1.73x$;

(c) £29 400, estimate of takings when no part-time staff employed. £173 estimate of extra taken per extra part-time member of staff;

(d) Week with 28 staff – abnormally low takings relative to pattern;

(e) Easier to organise steady increase of part-time staff than week-to-week fluctuations, however increasing number many coincide with another factor – such as run-up to Christmas – which will increase takings. From a statistical point of view, it is better to randomly order the weeks in which the given numbers of part-time staff are used.

11 (a) and **(b)**

Graph of estimated age against actual age

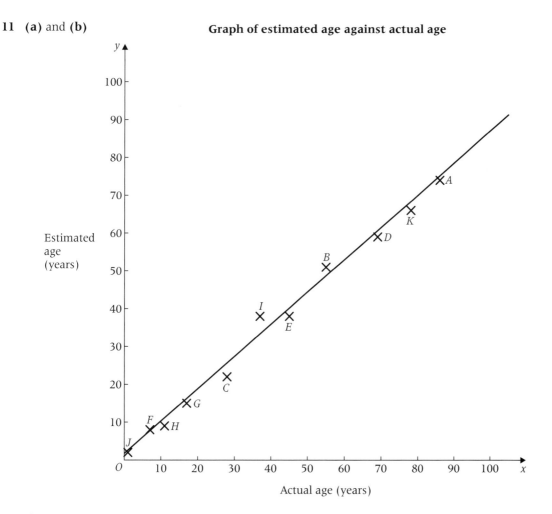

(b) $y = 0.984 + 0.853x$;

(c) $D - -0.860 \quad I - 5.44$;

(d) D has a small residual but is a poor estimate.
I has a large residual but is a good estimate.
Small residual indicates similar pattern to other estimates.
As Paulo tends to underestimate ages this means a poor estimate
in this case.

12 (a) and **(b)**

Graph of load time against hired help

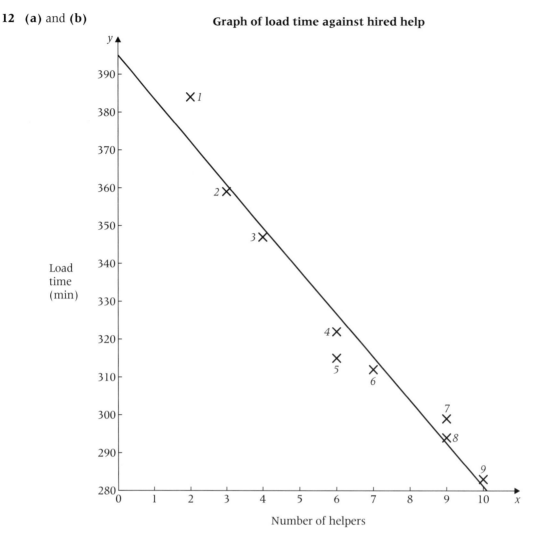

(b) $y = 396 - 11.5x$;

(c) a – estimate of time to load van with no local people hired
b – estimate of reduction in time to load van for each local
person hired;

(d) Extrapolation, estimate would be negative, which is impossible;

(e) Cannot tell if reduction in time is due to more people hired or to
roadies gaining more experience.

Exam style practice paper

1 (a) 0.289;   **(b)** 0.201;   **(c)** 0.290.

2 A parameter is a numerical property of a population, a statistic is a
numerical property of a sample.

3 (a) (i) 0.333,   **(ii)** 0.4,   **(iii)** 0.4;

(b) (i) W, X. Member cannot be both adult and junior,

(ii) $V, X \ \mathrm{P}(X) = \mathrm{P}(X \mid V)$.

4 **(a)** median 52.4 (52–53), interquartile range 20.7 (19–21);

(b) mean 54.2, sd 17.4, allow 17.1;

(c) 54.1 (53.5–54.5);

(d) (i) median – cannot calculate mean without knowing times of slowest three assembly workers (other reasons accepted),

(ii) 54.1.

5 **(a)** 36.3–60.5;

(b) (i) 46.42–52.58, **(ii)** 17.2–81.8;

(c) Evidence mean is greater than 25 but some individuals will be less than 25.

6 **(a)** and **(b)**

Graph of cut time against time since last cut

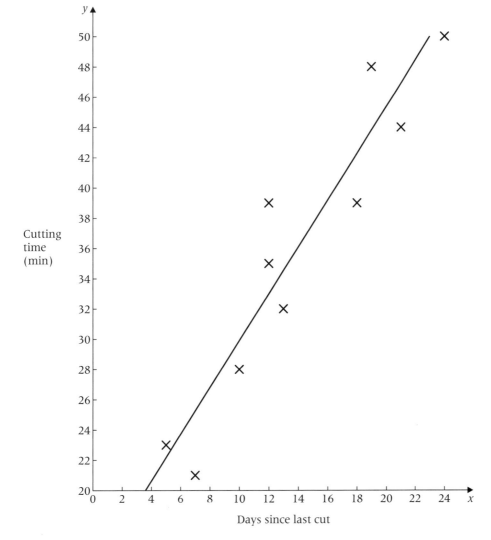

(b) $y = 14.6 + 1.51x$;

(c) (i) Estimate of additional time to cut lawn for each additional day since last cut,

(ii) Estimate of time to cut lawn a second time immediately after first cut.

    (d) Extrapolation unwise. Grass may be too long for lawnmower to function (or other sensible answer);

    (e) Time since last cut will affect time to cut. Time to cut will not affect time since last cut.

7 (a) (i) 0.977,   **(ii)** 0.819;

    (b) 8.20 a.m;

    (c) 8.17 a.m;

    (d) (i) Shortest mean travelling time,

        (ii) Least variable travelling time.

Index